Table Of Contents

Chapter 1: Introduction to Interplanetary Travel...............................3
Chapter 2: Overview ofThe Solar System...3
Chapter 3: Mercury:The Innermost Planet..3
Chapter 4: Venus: Earth's Twin..3
Chapter 5: Earth: The Home Planet..3
Chapter 6: Mars:The Red Planet..3
Chapter 7: The Asteroid Belt:A Treasure Trove 3
Chapter 8: Jupiter:The Gas Giant 3
Chapter 9: Saturn: The Ringed Wonder 3
Chapter 10: Uranus and Neptune:The Ice Giants.............................3
Chapter 11: Beyond the Solar System: The Milky Way Galaxy........3
Chapter 12: The Technologies of Interplanetary Travel 3
Chapter 13: The Future ofHuman Exploration.................................3
Chapter 14: Preparing for the Journey...3
Chapter 15: Conclusion:The Next Frontier 3
Saturn Cafe 1

Copyright © 2024 by Hakeem Ali-Bocas Alexander & UniquilibriuM, LLC Publishing.

All rights reserved.

No portion of this book may be reproduced in any form without written permission from the publisher or author, except as permitted by U.S. copyright law.

This publication is designed to provide accurate and authoritative information in regard to the subject matter covered. While the publisher and author have used their best efforts in preparing this book, they make no representations or warranties with respect to the accuracy or completeness of the contents of this book and specifically disclaim any implied warranties of merchantability or fitness for a particular purpose. The advice and strategies contained herein may not be suitable for your situation. You should consult with a professional when appropriate. Neither the publisher nor the author shall be liable for any loss of profit or any other commercial damages, including but not limited to special, incidental, consequential, personal, or other damages.

Saturn Cafe

Mass collaboration is exceedingly urgent now. I want to lunch while watching rings zoom by at 40,000 mph at the Saturn Cafe. I want to meet blue girls from Andromeda next door.

I want to catch a ride with a Graviton to the next Brane and visit my inter-dimensional cousins, and I probably will if I already haven't (I have the best lucid dreams).

While looking through telescopes and microscopes for centuries, pondering its mysteries and speculating the meaning of it all through mathematical tomes; we have uncovered a few secrets of this Universe. Nano-Tech and nuclear fusion are but a few of these inspirations.

Hermes Trismegistos etched upon his Emerald Tablet "…that which is above is as that which is below, and that which is below is as that which is above, to perform the miracles of the One Thing ."

Likewise, we discover and invent as we peer into the micro-cosmos and amongst the stars.

Excerpt from "**Exercising more than your mind…**" by Hakeem Ali-Bocas Alexander,
March 24th 2006 (https://eym.hypnoathletics.com/434/)

Images of planets and NASA quotes were sourced from
NASA Science Space Place (https://spaceplace.nasa.gov/); &
NASA Science (https://science.nasa.gov/)

Other images were generated by **NOVA AI** powered by **Chat GPT** with other design enhancements added using **Canva** and **Iconfinder**.

All of it is a vision I share with others for the future of humanity.

The Dawn of Space Exploration

The dawn of space exploration marked a pivotal moment in human history, characterized by the relentless pursuit of knowledge beyond our earthly confines. Following centuries of astronomical observation and theoretical speculation, the mid-20th century emerged as a catalyst for this new era. The Cold War rivalry between the United States and the Soviet Union ignited a fervent race for technological supremacy, propelling humanity into the cosmos. This period saw the development of groundbreaking technologies and the establishment of space agencies dedicated to exploring the final frontier.

The launching of by the Soviet Union in 1957 is widely regarded as the moment when humanity first stepped into the realm of space exploration. This historic satellite, the first artificial object to orbit Earth, not only demonstrated the feasibility of space travel but also captured the imagination of people worldwide. The subsequent launch of by the United States in 1958 further solidified the need for a coordinated effort in space research. These early missions laid the groundwork for future exploration and inspired generations to look skyward, envisioning a future where interplanetary travel would become a reality.

As the 1960s unfolded, the space race intensified, culminating in monumental achievements such as the Apollo program. **The Apollo 11** mission in 1969, which successfully landed humans on the , showcased the extraordinary capabilities of human ingenuity and engineering. and 's historic steps on the lunar surface not only fulfilled a national goal but also opened the door to broader cosmic ambitions. This monumental event served as a catalyst for scientific inquiry, igniting interest in the potential for life beyond Earth and the exploration of other celestial bodies within our solar system.

With the Moon landing as a springboard, the scope of space exploration expanded significantly in the following decades. Missions such as the spacecraft, launched in 1977, ventured far beyond the inner solar system, providing unprecedented data about the outer planets and their moons. These robotic emissaries enabled scientists to glean insights into the composition and dynamics of our celestial neighbors, reshaping our understanding of planetary science. The success of these missions demonstrated the viability of long-duration space exploration and paved the way for future endeavors that would seek to reveal the mysteries of the Milky Way Galaxy.

As we stand on the threshold of a new era in space exploration, the lessons learned from the early days serve as a foundation for current and future missions. The advent of advanced technologies, international collaborations, and an increasing interest in commercial spaceflight signal a vibrant future for interplanetary travel. The exploration of , the search for extraterrestrial life, and the potential for human settlement on other planets are no longer distant dreams, but attainable goals. As explorers, we are called to embrace the spirit of discovery that ignited the dawn of space exploration, propelling humanity toward a future filled with infinite possibilities among the stars.

The Vision of Interplanetary Travel

The vision of interplanetary travel has evolved significantly over the centuries, transitioning from mere speculation to a tangible goal within reach. Early science fiction depicted journeys to other planets as fantastical adventures fueled by imagination. However, contemporary advancements in technology and science are transforming these dreams into actionable plans. The idea of traveling beyond Earth is no longer confined to the pages of novels; it is being meticulously mapped out by scientists, engineers, and visionaries dedicated to exploring our solar system.

At the heart of interplanetary travel is the ambition to establish human presence beyond Earth. One thing this vision encompasses is visiting other celestial bodies where sustainable habitats that could support life are established. Mars, with its Earth-like qualities, remains the most prominent candidate for human colonization. Missions like NASA's program aim to return humans to the Moon, serving as a stepping stone for deeper space exploration and a testing ground for technologies necessary for Mars missions. Each successful mission paves the way for future explorers to consider the possibility of living and working on another planet.

The technological advancements required for interplanetary travel are formidable, yet the potential benefits are immense. Innovations in propulsion systems, life support technologies, and sustainable resource utilization are critical for long-duration space missions. The development of ion propulsion and nuclear thermal propulsion systems could reduce travel time to Mars and beyond, making interplanetary journeys more feasible. Additionally, the ability to produce food and water from Martian resources could enhance the viability of human settlements, ensuring that explorers can thrive in environments previously deemed inhospitable.

The vision of interplanetary travel extends beyond individual missions; it encompasses the creation of a broader framework for exploring the Milky Way Galaxy. This idea of a "*solar system passport*" captures the essence of this exploration, allowing humanity to chart its course through space. International collaboration is of course essential for this endeavor, as sharing knowledge and resources can accelerate progress. With mutually beneficial partnerships among nations and private enterprises, the dream of a multi-planetary civilization can become a reality, inspiring future generations to explore the cosmos.

For many visionaries, the dream of interplanetary travel is about much more than reaching new destinations; it is also about expanding the horizons of human experience and understanding. As explorers, the quest for knowledge drives us to seek answers to fundamental questions about our existence and the universe. Each mission to another planet is a step toward unraveling the mysteries of the cosmos, inviting us to ponder our place within it. The journey beyond the blue planet is an invitation to embrace our curiosity and to boldly venture into the unknown, transforming the vision of interplanetary travel into a shared legacy for all of humanity.

Tools of the Trade

On the quest for interplanetary travel and exploration, certain tools of the trade are essential for successful missions beyond Earth's atmosphere. As explorers set their sights on distant planets and moons within our solar system, they rely on a range of sophisticated instruments designed to gather data, perform analysis, and ensure safe travel. These tools encompass everything from advanced spacecraft technology to scientific instruments capable of withstanding the harsh conditions of space.

Spacecraft serve as the primary vessels for exploration, equipped with state-of-the-art propulsion systems that enable them to traverse vast distances. Ion thrusters, for example, utilize electric fields to accelerate ions, providing efficient propulsion for long-duration missions. Additionally, traditional rocket engines continue to play a pivotal role, allowing for rapid launches and maneuvers. The design of these spacecraft must also consider the challenges of radiation, temperature extremes, and micrometeoroid impacts, making robust engineering a critical component of their development.

Scientific instruments aboard these spacecraft are invaluable for gathering data once explorers reach their destinations. Spectrometers and cameras are commonly used to analyze the chemical composition of planetary surfaces and atmospheres. These tools allow scientists to study everything from mineral deposits on Mars to the gaseous plumes of , a moon of . Remote sensing technology also plays a crucial role, enabling explorers to collect data from afar without the need for direct contact with potentially hazardous environments.

Communication systems stand as another vital tool in the interplanetary toolkit. Given the immense distances involved in solar system exploration, maintaining contact with mission control on Earth can be challenging. High-gain antennas and relay satellites help ensure that data is transmitted back to Earth efficiently. The delay in signals due to the vast distances also necessitates autonomous systems that allow spacecraft to make real-time decisions based on the data they collect, further enhancing the exploration capabilities of these missions.

And of course, the human element of exploration must be given the highest priority. Astronauts and robotic operators are equipped with specialized training and tools to respond to unexpected challenges in space. From advanced life support systems that provide breathable air and water to extravehicular activity suits that protect against the vacuum of space, these tools are designed to ensure the safety and success of human explorers. As technology continues to evolve, the tools of the trade will undoubtedly expand, charting a course for deeper exploration and greater understanding of our solar system and beyond.

The Formation of the Solar System

The formation of the solar system is a fascinating tale that began approximately 4.6 billion years ago. It all started with a vast cloud of gas and dust, known as a solar nebula, which was primarily composed of hydrogen and helium, along with heavier elements produced by previous generations of stars. This nebula existed in the Milky Way galaxy, swirling and coalescing under the influence of gravity. Over time, disturbances, possibly from nearby supernovae, triggered the collapse of this cloud, leading to the formation of the sun at its center.

As the solar nebula collapsed, it began to spin and flatten into a rotating disk. The gravitational forces within the disk caused particles to collide and stick together, forming larger and larger bodies through a process known as accretion. In the inner region of the disk, temperatures were high enough to vaporize lighter elements, leading to the formation of rocky planets. Here, materials such as silicates and metals combined to form the terrestrial planets: , , Earth, and Mars. These planets, characterized by their solid surfaces, emerged from the remnants of the primordial material.

In contrast, the outer regions of the solar system experienced cooler temperatures, allowing volatile compounds like water, ammonia, and methane to remain in solid form. This enabled the formation of gas giants such as and Saturn, which accumulated thick atmospheres of hydrogen and helium. Meanwhile, the icy bodies beyond , known as the , formed smaller icy planets and dwarf planets like . These processes created a diverse array of celestial bodies, each with unique characteristics and compositions, which together make up our solar system.

The gravitational interactions among these newly formed bodies played a crucial role in shaping the solar system as we know it today. As planets formed, their orbits were not perfectly circular; instead, they were influenced by the gravitational pull of neighboring bodies. This led to a dynamic environment where collisions and interactions occurred frequently. The result is that many of these early planetesimals were either ejected from the solar system or became part of larger bodies, resulting in the relatively stable configuration of planets and moons we observe in the present day.

The study of the solar system's formation provides insights into our own cosmic neighborhood and also offers clues about the processes that govern planetary systems throughout the . Continued research of these mechanisms can inform future interplanetary exploration, guiding explorers as they seek to uncover the mysteries of other solar systems. When we dare to venture beyond our blue planet, the knowledge gained from our own solar system's origins will be instrumental in navigating the vast expanse of space, offering a passport to new frontiers and discoveries waiting to be made.

The Structure of the Solar System

The structure of the solar system is a fascinating and intricate arrangement of celestial bodies that extends far beyond the familiar planets we observe from Earth. At its center lies the , a massive ball of plasma that provides the necessary heat and light for life on our planet. The Sun accounts for more than 99% of the total mass of the solar system, exerting a gravitational pull that governs the orbits of all other objects within its reach. Surrounding the Sun are eight major planets, each unique in its composition, atmosphere, and potential for exploration.

The major planets can be divided into two distinct categories: terrestrial planets and gas giants. —Mercury, Venus, Earth, and Mars—are characterized by their solid, rocky surfaces and relatively smaller sizes. They are positioned closer to the Sun and have varying atmospheres, with Earth being the only one known to support life.

Beyond Mars lies the asteroid belt, a region filled with rocky debris that serves as a boundary between the inner and outer solar system. This belt is of particular interest to explorers, as it contains remnants from the solar system's formation.

 is dominated by the gas giants: Jupiter, Saturn, , and Neptune. These planets are significantly larger than their terrestrial counterparts and are composed primarily of hydrogen and helium, with thick atmospheres and no solid surface. Jupiter, the largest planet, is known for its , a massive storm that has been raging for centuries. Saturn, with its iconic rings, presents unique opportunities for study regarding planetary formation and the dynamics of ring systems. The gas giants not only captivate the imagination but also harbor numerous moons, some of which, like and , are considered prime candidates for astrobiological research.

Beyond the gas giants lies the Kuiper Belt, a vast region filled with icy bodies, dwarf planets like Pluto, and other small celestial objects. This belt marks the transition from the main solar system to the more distant areas of the solar system, including the scattered disk and the . The Kuiper Belt is of great interest to explorers as it contains primitive material that can offer insights into the early solar system and the processes that shaped it. Missions to this region, such as the 's flyby of Pluto, have opened up new avenues for discovery and understanding.

The structure of the solar system is not static; it is constantly evolving due to gravitational interactions, collisions, and the influence of external forces. Keeping an eye on this dynamic system is crucial for future interplanetary travel and exploration. When explorers venture beyond Earth, the knowledge of the solar system's structure will guide their missions, helping them navigate the complexities of celestial mechanics while seeking to uncover the mysteries that lie within our cosmic neighborhood. Each journey into the solar system is an opportunity to expand our understanding of the universe and our place within it.

The Role of the Sun

The Sun stands as the central figure of our solar system, a massive ball of hydrogen and helium that exerts a gravitational hold on all celestial bodies within its reach. With a , it accounts for about . This immense gravitational force is what keeps planets, asteroids, and comets in stable orbits, allowing for the complex dance of celestial mechanics that explorers seek to understand. The Sun is not only a source of light and heat but also a driving force behind the dynamic processes that govern the solar system's structure and evolution.

The process of at the Sun's core produces energy that radiates outward, providing the warmth necessary for life on Earth. This energy influences planetary atmospheres, climates, and the potential for habitability on various worlds. Explorers venturing beyond Earth must consider how impacts their missions. For example, the radiation environment varies significantly between planets based on their distance from the Sun and their . Of course, engineering with respect to these factors is foundational for designing spacecraft and planning safe interplanetary travel.

The Sun also plays a vital role in shaping , streams of charged particles that flow outward into space. These solar winds interact with , creating phenomena such as and influencing space weather. For explorers, monitoring solar activity is essential, as and can pose significant risks to spacecraft and astronauts. Awareness of these solar phenomena allows for better preparedness and risk mitigation during missions, especially for those venturing to destinations like Mars or the outer planets.

In addition to its physical influence, the Sun serves as a key marker for navigation within the solar system. Its position and brightness provide a reliable point of reference for spacecraft, helping explorers chart their courses through the vastness of space. The Sun's cycles, including its of activity, can also impact long-term planning for exploration missions, as periods of heightened solar activity can affect communication and navigation systems. Tracking and recording these cycles provides critical data for maintaining the safety and efficiency of interplanetary journeys.

While we humans sometimes use the Sun as a backdrop for exploration; we also know that it is integral to understanding the broader context of our galactic neighborhood. Its gravitational influence extends into the Milky Way, affecting the orbits of nearby stars and contributing to the overall dynamics of our galaxy. For explorers looking to venture beyond the confines of the solar system, knowledge of the Sun's role within the larger framework of galactic exploration is exciting and essential. This understanding emphasizes the interconnectedness of cosmic phenomena and highlights the importance of the Sun as both a life-giving star and a key player in the grand scheme of the universe.

Characteristics and Composition

According to NASA: "".

Starting with the terrestrial planets, Mercury, Venus, Earth, and Mars share a rocky composition, yet they differ significantly in their atmospheres and surface conditions. Mercury, closest to the Sun, has a thin atmosphere and extreme temperature fluctuations, making it a challenging destination for exploration Venus, shrouded in thick clouds of sulfuric acid, presents an inhospitable environment with crushing pressures and high temperatures. Earth, our home, boasts a diverse range of environments and is the only known planet to support life. Mars, with its evidence of past water flows and potential for microbial life, remains a primary target for exploration efforts.

Studying the characteristics and composition of celestial bodies within our solar system offer essential insights for our understanding of their formation, evolution, and potential for exploration. Each planet, moon, asteroid, and comet exhibits distinct features that inform scientists about the conditions prevalent during their creation and the processes that have shaped them over billions of years. These traits can include size, mass, density, atmospheric conditions, surface geology, and magnetic fields, all of which are essential for explorers planning interplanetary missions.

The gas giants—Jupiter and Saturn—along with the ice giants, Uranus and Neptune, present a contrasting composition. These planets are primarily composed of hydrogen and helium, with deep atmospheres and no solid surface. Jupiter, the largest planet in our solar system, is known for its Great Red Spot, a massive storm, and its extensive system of moons, including the ocean world Europa, which may harbor conditions suitable for life. Saturn, famed for its striking rings, also has a rich array of moons, such as Titan, which features lakes of liquid methane and an atmosphere denser than Earth's.

Beyond the major planets lie numerous dwarf planets and small bodies, including asteroids and comets. The asteroid belt, located between Mars and Jupiter, is home to a vast number of rocky fragments that provide insights into the early solar system. Comets, originating from the Kuiper Belt and Oort Cloud, are icy bodies that develop spectacular tails when they approach the Sun. These small celestial bodies are not only of scientific interest but also potential resource targets for future missions, providing materials for fuel and construction in space.

Continued research and documentation the characteristics and composition of these celestial entities is fundamental for explorers seeking to navigate the complexities of interplanetary travel. Knowledge of surface conditions, gravitational forces, and potential hazards allows for the design of appropriate spacecraft and mission strategies. During our adventures beyond our blue planet, each journey into the solar system promises to unveil secrets of the cosmos, inspiring future generations of explorers to push the boundaries of human knowledge and experience.

Historical Missions

Historical missions to other planets in the Solar System mark significant milestones in humanity's quest for knowledge and exploration. Beginning in the 1960s, the space race catalyzed a series of groundbreaking missions that expanded our understanding of the celestial bodies within our reach. The Soviet Union's **Luna program**, which successfully sent robotic landers to the Moon, set the stage for subsequent explorations, culminating in the Apollo missions that enabled humans to walk on the lunar surface. These early efforts ignited public interest and laid the groundwork for future interplanetary missions.

NASA's **Mariner program** was pivotal in the exploration of Mars and Venus. **Mariner 4**, which flew past Mars in 1965, provided the first close-up images of the Martian surface, revealing a cratered landscape that hinted at a more complex geological history. Following this, Mariner 9 became the first spacecraft to orbit another planet, capturing detailed images of Martian features such as volcanoes and canyons. These missions not only advanced our understanding of planetary atmospheres and geology but also inspired the possibility of future human exploration.

The Voyager missions, launched in 1977, marked a new era in interplanetary exploration. Voyager 1 and **Voyager 2** conducted flybys of the outer planets, including Jupiter, Saturn, Uranus, and Neptune, providing unprecedented data about their atmospheres, moons, and ring systems. The iconic images of the "Pale Blue Dot," taken by Voyager 1 from a distance of over three billion miles, served as a poignant reminder of Earth's place in the universe. The Voyager spacecraft continue to send back data as they journey into interstellar space, symbolizing humanity's enduring quest to understand the cosmos.

Mars missions have been a focal point of exploration, with several successful landers and rovers enhancing our knowledge of the Red Planet. NASA's **Pathfinder mission** in 1997 introduced the **Sojourner rover**, which was the first to explore Mars' surface. This was followed by the **Spirit** and **Opportunity** rovers, which provided evidence of past water activity on Mars. The **Curiosity** rover, launched in 2011, has been instrumental in studying the planet's habitability, analyzing soil samples, and searching for organic compounds. Each mission builds on the last, curating a scientific archive of planetary exploration that informs future endeavors.

As we look ahead, historical missions serve as a foundation for upcoming explorations, including planned missions to the Moon and Mars by NASA's Artemis program and the Mars Sample Return mission. These initiatives have the aim of revisiting our closest celestial neighbors and preparing for eventual human settlement beyond Earth. The legacy of past missions is critical in shaping our strategies for interplanetary travel, emphasizing the importance of learning from previous successes and challenges. Explorers who venture into the cosmos are fortunate to be able to carry with them the knowledge and determination forged through decades of historical missions.

Future Exploration Plans

The future of solar system exploration is poised to take a transformative leap forward, with ambitious plans that promise to unlock the secrets of our celestial neighborhood. Space agencies and private companies are investing heavily in technology and missions aimed at expanding humanity's reach beyond Earth. These endeavors are not limited to reconnaissance; they are designed to pave the way for permanent human presence on other planets, advanced scientific research, and the potential for interplanetary travel. Exploration plans include missions to the Moon, Mars, and beyond, each with unique objectives and challenges that will enhance our understanding of the universe.

One of the most immediate goals is the establishment of a sustainable presence on the Moon, which serves as a vital stepping stone for deeper space exploration. NASA's Artemis program aims to return humans to the lunar surface by the mid-2020s, focusing on the lunar South Pole. This region is of particular interest due to its potential water ice deposits, which could support future missions. The Moon's proximity to Earth makes it an ideal testing ground for new technologies and life support systems necessary for longer missions to Mars and beyond. International collaborations and partnerships with private space firms will play a crucial role in the success of these lunar endeavors.

Mars exploration remains a central focus for future missions, with plans for both robotic and crewed missions in the next few decades. NASA's [Perseverance rover](), launched in 2021, is currently exploring the Martian surface, collecting samples and searching for signs of ancient life. Following this, a series of missions are being planned to return these samples to Earth, further enriching our understanding of the Red Planet. Human missions to Mars are projected to occur in the 2030s, with private companies like SpaceX also working on their own initiatives to transport humans to the Martian surface. These missions will study Mars's geology, climate, and test the viability of life support technologies and habitats for future colonization.

The exploration of asteroids and the outer planets is also on the horizon, with missions designed to investigate these celestial bodies for scientific and resource potential. NASA's **Psyche mission**, set to launch in the mid-2020s, will explore a unique metallic asteroid believed to be the core of a protoplanet. Similarly, missions to the moons of Jupiter and Saturn, such as **Europa Clipper** and **Dragonfly**, aim to investigate their potential for harboring life. These missions will provide insights into the formation of the solar system and the conditions necessary for life beyond Earth, expanding our understanding of where we fit within the Milky Way Galaxy.

As we look further into the future, concepts for interstellar travel are beginning to take shape, albeit in theoretical stages. Projects like **Breakthrough Starshot** envision sending small, light-propelled spacecraft to nearby star systems within a human lifetime. While these concepts are in the early stages, they inspire a vision of humanity's eventual expansion beyond the solar system. The ongoing advancements in propulsion technologies, robotics, and artificial intelligence will be crucial in making these ambitious plans a reality. As explorers, we stand on the brink of a new era in space exploration, with the promise of discovery and the potential to become an interstellar civilization within our grasp.

Chapter 4: Venus: Earth's Twin

Atmospheric Conditions

Atmospheric conditions are of primary influence in the exploration of our Solar System, effecting everything from spacecraft design to landing techniques and habitability potential on other celestial bodies. Each planet and moon presents unique atmospheric characteristics that can either facilitate or hinder exploration efforts. Being able to plan according to these conditions is essential for both current and future missions, as they inform the strategies and technologies necessary to navigate the diverse environments of the Solar System.

Earth, our home planet, serves as a reference point for atmospheric studies. It has a complex atmosphere composed primarily of nitrogen and oxygen, with weather patterns and climate systems that can change rapidly. These dynamics are well understood, allowing for sophisticated predictions and preparations for exploration missions. In contrast, other planets like Venus and Mars exhibit vastly different atmospheric features.

Venus has a thick, toxic atmosphere dominated by carbon dioxide, with surface pressures that are about 92 times that of Earth. This extreme environment poses significant challenges for landers and rovers, requiring specialized materials and engineering solutions to withstand the harsh conditions.

NASA tells us: "."

Mars, with its thin atmosphere, presents a different set of challenges and opportunities for exploration. Composed mostly of carbon dioxide, with trace amounts of nitrogen and argon, Mars' atmosphere is incapable of supporting human life without proper life support systems. However, its thinness allows for less atmospheric drag, making it easier for spacecraft to enter and land. Our familiarity with the Martian atmosphere is vital for assessing weather patterns, dust storms, and the potential for future human habitation. The exploration of Mars has revealed evidence of ancient water flows, suggesting that its atmosphere was once much thicker and could have supported microbial life.

Gas giants like Jupiter and Saturn offer a striking contrast with their vast and tumultuous atmospheres. These planets are predominantly composed of hydrogen and helium, featuring dynamic weather systems, including the famous Great Red Spot on Jupiter—a massive storm larger than Earth. The study of these atmospheric conditions is important not only for understanding planetary formation and evolution but also for preparing missions to their moons, such as Europa and Titan, which may harbor subsurface oceans and potentially life. The extreme pressures and temperatures present in the atmospheres of gas giants require innovative spacecraft designs and technologies to ensure mission success.

In addition to the major planets, moons like Titan and Enceladus have atmospheres that present unique exploration opportunities. Titan, Saturn's largest moon, boasts a dense atmosphere rich in nitrogen, with lakes of methane and ethane on its surface. This intriguing environment challenges our traditional views of habitability and opens up new avenues for scientific inquiry. Enceladus, on the other hand, has a tenuous atmosphere but is famous for its geysers that spew water vapor and organic molecules into space, hinting at the possibility of life in its subsurface ocean. For explorers venturing beyond the confines of Earth, understanding atmospheric conditions across the Solar System will be fundamental to unlocking the mysteries of these celestial bodies and the potential for interplanetary travel and exploration in the Milky Way Galaxy.

Surface Features and Geography

The surface features and geography of celestial bodies within our solar system offer a rich diversity of landscapes that reflect their unique histories and the processes that have shaped them. From the towering mountains of Mars to the icy plains of Europa, each planet and moon presents a distinct environment ripe for exploration. Observation of these features enhances our appreciation of these worlds, while informing the strategies we will employ in our interplanetary travels. As explorers, we must familiarize ourselves with the geographical nuances that could influence our missions.

Mars, often regarded as the most Earth-like planet, boasts a diverse array of surface features. Its landscape is marked by the largest volcano in the solar system, , which stands nearly three times the height of . The planet also hosts the vast canyon system of , a chasm that stretches over . These geological marvels are visually striking, key indicators of Mars' geological history and potential for past water activity. Researching how these features came to be formed will provide essential insights for any future human settlement or exploration efforts.

In contrast, the gas giants, such as Jupiter and Saturn, present a different set of geographical characteristics. While their solid surfaces remain enigmatic, their swirling atmospheres are filled with dynamic weather patterns and storms. Jupiter's Great Red Spot, a massive storm larger than Earth, has raged for centuries. The study of these atmospheric phenomena is essential for understanding the planets' meteorological systems and potential for hosting weather-related phenomena that could impact exploration. Another point of interest are the beautiful and intricate ring systems of Saturn that offer a glimpse into the processes of planetary formation and the gravitational interactions within the solar system.

The icy moons of the outer planets, particularly Europa and Enceladus, captivate explorers with their potential for harboring life. Europa's surface is characterized by a shell of ice that conceals a vast ocean beneath, while Enceladus is known for its geysers that eject plumes of water vapor and organic compounds into space. These surface features suggest the presence of liquid water—an essential ingredient for life—and provide tantalizing clues about the moons' geologic activity. Carefully mapping the geography of these moons is integral as we plan missions that aim to probe their subsurface oceans and search for signs of extraterrestrial life.

The diverse landscapes of asteroids and comets also play a vital role in our understanding of the solar system's formation. These small celestial bodies exhibit features such as craters, ridges, and, in the case of comets, spectacular tails formed by the sublimation of ice. Studying these surface features can reveal the conditions of the early solar system and the processes that led to the formation of planets. Explorers who venture into these regions will undoubtedly encounter the thrill of discovery, and the potential for resource utilization, which could be pivotal for sustaining long-term missions deeper into the Milky Way Galaxy.

In preparation for interplanetary travel, a thorough understanding of surface features and geographical contexts will prove invaluable. Each celestial body presents unique challenges and opportunities that will shape our exploration strategies. From the rugged terrains of Mars to the icy expanses of Europa, these geographical elements will inform our technological advancements and mission planning. Through continued scrutiny of the diverse landscapes of our solar system, we equip ourselves with the knowledge necessary to navigate and explore the cosmos, leading to the expansion of humanity's presence beyond the confines of Earth.

Prospects for Exploration

The prospects for exploration within our solar system present an exciting frontier for humanity, as advances in technology and a growing understanding of celestial bodies pave the way for future missions. The Moon, Mars, and the moons of Jupiter and Saturn have emerged as prime targets for exploration. Each of these destinations offers unique opportunities for scientific discovery, resource utilization, and potential human settlement. With initiatives like NASA's Artemis program aiming to return humans to the Moon and missions like Mars 2020 and Artemis I setting the stage for deeper exploration, the solar system is on the brink of a new era.

The Moon serves as a critical stepping stone for interplanetary travel, acting as a testing ground for technologies and strategies that can be applied to more distant missions. The presence of water ice in permanently shadowed craters could support future human outposts and provide resources for fuel and life support. By establishing a sustained human presence on the Moon, explorers can gather invaluable data on the effects of long-term space habitation, which is essential for future missions to Mars and beyond. The potential for lunar mining also opens avenues for extracting materials that could be used in building infrastructure for deeper space exploration.

Mars remains the most compelling target for human exploration, with its similarities to Earth and the tantalizing possibility of past or present life. Robotic missions have already laid the groundwork for future manned missions by identifying key resources such as water and analyzing the planet's geology. As technology advances, plans for human landings on Mars are becoming more concrete, with timelines ranging from the 2030s onward. The challenges of the Martian environment, including radiation exposure and extreme temperatures, require innovative solutions, making exploration of Mars a scientific endeavor that will push the limits of human ingenuity and resilience.

The gas giants and their moons, particularly Europa and Enceladus, continue to inspire advancements in exploration. These moons are believed to harbor subsurface oceans beneath their icy crusts, offering potential habitats for extraterrestrial life. Missions like NASA's Europa Clipper aim to study these environments from orbit, while future lander missions could directly analyze the water and potential biosignatures. Getting closer to these distant worlds could revolutionize our knowledge of life in the universe and our place within it. The prospect of exploring these moons contributes to a broader understanding of planetary systems and the conditions that support life.

As we look beyond our solar system, the Milky Way Galaxy itself beckons with the promise of discovery. Advances in propulsion technologies, such as ion drives and theoretical concepts like warp drives, could eventually enable interstellar travel. The search for exoplanets in habitable zones around other stars opens new possibilities for exploration and potential colonization. Initiatives like the Breakthrough Starshot project aim to send small spacecraft to the nearest star systems, exploring the potential for life beyond our solar neighborhood. The prospects for exploration are not limited to our immediate celestial companions; they extend into the vastness of space, urging us to dream and innovate in pursuit of the unknown.

Unique Features of Earth

Earth, the third planet from the Sun, possesses a remarkable array of unique features that set it apart within our solar system. One of the most striking aspects of Earth is its abundant liquid water, which covers approximately 71% of its surface. This vast expanse of oceans not only plays a lead role in regulating the planet's climate but also supports an incredible diversity of life forms. The presence of water in its three states—liquid, solid, and gas—creates a dynamic environment that links ecosystems and sustains biological processes essential for life.

Another defining characteristic of Earth is its protective atmosphere, composed primarily of nitrogen and oxygen. This layer shields the planet from harmful solar radiation and cosmic rays, allowing life to thrive on its surface. The atmosphere also facilitates the greenhouse effect, which retains heat and maintains a stable climate. This delicate balance has enabled Earth to support a wide range of habitats, from the frozen tundras to lush rainforests, showcasing the planet's biodiversity and ecological complexity.

Earth's geological activity is another feature that distinguishes it from other planets. The presence of tectonic plates leads to earthquakes, volcanic eruptions, and the formation of mountains, constantly reshaping the planet's surface over time. This geological dynamism creates diverse landscapes while playing a vital role in nutrient cycling and carbon storage, influencing both climate and life. The ongoing processes of erosion and sedimentation further contribute to the varying states of Earth's environments.

The planet's magnetic field, generated by the motion of molten iron in its outer core, is yet another unique attribute. This magnetic field acts as a shield against solar winds and cosmic radiation, protecting life from harmful particles. It also plays a crucial role in navigation for both humans and many animal species. The phenomenon of auroras, which occur when charged particles from the solar wind interact with the atmosphere, adds a captivating visual element to Earth's magnetic properties.

Earth is distinguished by its position in the solar system, providing a unique vantage point for exploration. Its relative proximity to the Moon has made it an ideal launch point for human space travel, enabling missions that have expanded our understanding of the cosmos. Additionally, the presence of diverse ecosystems and the ability to support intelligent life make Earth a focal point in the quest for knowledge about the universe. As explorers venture beyond our blue planet, the unique features of Earth serve as a reminder of our home's significance in the broader context of the Milky Way Galaxy.

The Importance of Earth in the Solar System

The Earth, as the third planet from the Sun, holds a unique position within the Solar System. Its location in the habitable zone, often referred to as the "," allows for the presence of liquid water, a critical element for life as we know it. This specific distance from the Sun ensures that temperatures are neither too hot nor too cold, resulting in a diverse range of ecosystems. As we have come to know from exploring beyond our planet, understanding Earth's unique role in the Solar System has helped us to appreciate the delicate balance that sustains life.

Earth's atmosphere is another crucial factor contributing to its importance. Composed of nitrogen, oxygen, carbon dioxide, and other trace gases, this protective layer shields the planet from harmful solar radiation and space debris. Additionally, the atmosphere plays a vital role in regulating temperature and climate, making conditions suitable for life. Earth's unique features remind us of the challenges that other celestial bodies face, such as Mars and Venus, where atmospheric conditions diverge significantly from those on Earth. In our quest to understand the potential for life elsewhere, Earth's atmosphere provides a benchmark for comparison.

The geological diversity of Earth further enhances its significance in the Solar System. With its vast oceans, towering mountains, and varied landscapes, our planet showcases the dynamic processes that shape celestial bodies. The study of Earth's geology not only informs us about its history and the evolution of life but also provides insights into the geological processes that may occur on other planets and moons. By examining Earth's unique features, explorers can better comprehend the potential for similar geological activity elsewhere, particularly on bodies like Europa and Enceladus, which harbor subsurface oceans.

Earth's biodiversity is a testament to the intricate relationships between various life forms and their environments. This incredible diversity of life results from millions of years of evolution, adapting to a variety of ecological niches. The study of Earth's flora and fauna has shed light on the resilience of life while also being a model for what explorers might encounter on exoplanets or moons with the potential for habitability. The complexity of Earth's ecosystems equips explorers with the knowledge needed to search for and recognize signs of life in other corners of the Solar System.

Earth's solitary position as a supporter of what we know of life serves as a launch point for interplanetary exploration, making it the hub of human activity in the Solar System. The advancements in technology and scientific knowledge developed on Earth have propelled missions to Mars, the outer planets, and beyond. As explorers prepare to embark on journeys to discover new worlds, Earth's role as the cradle of human ingenuity and curiosity cannot be overstated. The experiences and lessons learned from our home planet will guide future endeavors, ensuring that the quest for knowledge continues as we seek to uncover the mysteries of the cosmos.

Earth in the Context of Interplanetary Travel

Earth serves as the launchpad for humanity's ambitious journey into the cosmos, and understanding its role in the context of interplanetary travel is essential for explorers venturing beyond our blue planet. As the cradle of human civilization, Earth possesses unique characteristics that facilitate space travel, such as its abundant resources, diverse ecosystems, and established technological infrastructure. These elements create a supportive environment for the development of interplanetary missions, making Earth not only a home but also a vital hub for future explorations.

The planet's atmosphere is a protective shield against cosmic radiation and meteorite impacts, which poses significant challenges for space travel. This natural barrier allows scientists and engineers to develop and test spacecraft in conditions that approximate the vacuum of space without the full exposure to the hazards found beyond. Earth's gravitational pull is also crucial, as it enables spacecraft to launch with significantly less energy compared to launching from other celestial bodies. This gravitational advantage allows for more efficient fuel use, enabling longer and more ambitious missions across the solar system.

In addition to its physical attributes, Earth is a center of human knowledge and innovation. The collaboration of scientists, engineers, and explorers has led to unprecedented advancements in technology that are vital for interplanetary travel. From rocket propulsion systems to life support technologies, the tools developed on Earth are tailored to meet the challenges of living and working in space. The ongoing research in fields such as astrobiology and environmental science further enhances our understanding of potential extraterrestrial environments, thus preparing explorers for the conditions they may encounter on other planets.

Earth's diverse ecosystems provide a valuable reference point for understanding potential alien habitats. By studying Earth's varied climates and biological systems, scientists gain insights into how life might adapt to different planetary conditions. This comparative analysis is crucial for missions aimed at exploring potentially habitable zones on Mars, Europa, and beyond. The knowledge gained from these studies informs the design of missions, ensuring that explorers are equipped to gather valuable data about the environments they will encounter.

In humanity's preparation for an era of interplanetary travel, the existence of our blue planet seems miraculous. For explorers, understanding the interplay between Earth and interplanetary travel is not just about appreciating our home planet; it is about recognizing the pivotal role it will continue to play in humanity's quest to explore the stars. It is the birthplace of our species and the foundation upon which we build our aspirations for exploration. Every mission launched into the solar system is a testament to Earth's role as a hub of innovation, collaboration, and preparation.

Mars' Geology and Atmosphere

From NASA we learn: "*Mars is a cold, bleak wasteland, with very thin air that we Earthlings could never breathe. However, many of the pictures our telescopes, orbiters, and rovers have sent back show signs that liquid water might have been on the surface of Mars long ago. All these signs of water are very exciting. Why? Because on Earth, almost everywhere there is water, there is life. Whether the water is boiling hot or frozen, some sort of creature seems to thrive in it*".

Mars, often referred to as the Red Planet, boasts a diverse and intriguing geology that captivates explorers eager to understand its complex history. The surface of Mars is marked by an array of geological features, including the largest volcano in the solar system, Olympus Mons, which stands nearly three times the height of Mount Everest. Its broad shield shape indicates a long history of volcanic activity, suggesting that the planet was once geologically active. In addition to volcanoes, Mars is home to Valles Marineris, a canyon system that dwarfs the Grand Canyon, stretching over 4,000 kilometers in length. These features provide key insights into the tectonic and climatic processes that have shaped Mars over billions of years.

The Martian landscape is also dotted with ancient river valleys and lake beds, hinting at a wetter past when liquid water may have flowed freely across the surface. The presence of minerals such as clays and sulfates indicates that water played a significant role in Mars' geological evolution. Recent missions have discovered evidence of recurring slope lineae, dark streaks that appear and disappear with the seasons, suggesting the presence of liquid brines. This evidence of past and possibly present water encourages optimism about the planet's potential to support life and informs the ongoing search for biosignatures.

Mars' atmosphere is thin compared to Earth's, composed primarily of carbon dioxide (about 95.3%), with traces of nitrogen and argon. This tenuous atmosphere has significant implications for exploration and habitability. At an average pressure of less than one percent of Earth's, the Martian atmosphere is insufficient to support human life without adequate protection. Dust storms are a common phenomenon on Mars, capable of enveloping the entire planet and affecting solar-powered missions. Engineering for the dynamics of these storms and the overall atmospheric conditions on Mars is critical for future human exploration and habitation.

The atmospheric conditions on Mars contribute to its distinctive reddish appearance, which is due to iron oxide, or rust, on its surface. The planet experiences temperature extremes, with daytime temperatures near the equator reaching up to 20 degrees Celsius, while nighttime temperatures can plunge to minus 73 degrees Celsius. These variations are exacerbated by the thin atmosphere, which cannot retain heat effectively. For explorers, these climatic challenges highlight the need for advanced technology and planning to ensure the safety and success of missions.

As we continue to study Mars' geology and atmosphere, we unlock the secrets of its past and its potential for future exploration. The data gathered from rovers and orbiters enhances our understanding of the Martian environment, informing the design of habitats and life-support systems for human explorers. With each mission, we draw closer to revealing the mysteries of the Red Planet, laying the groundwork for future journeys in our quest to explore the solar system and beyond.

Robotic Exploration of Mars

Robotic exploration of Mars has been a cornerstone of our understanding of the Red Planet, marking significant milestones in the history of space exploration. Since the 1960s, a series of robotic missions have been launched to gather data about Mars' atmosphere, geology, and potential for life. These missions have employed various types of spacecraft, including orbiters, landers, and rovers, each designed to fulfill specific scientific objectives. The wealth of information amassed from these explorations has certainly deepened our knowledge of Mars in addition to laying the groundwork for future human missions.

One of the most notable robotic missions was NASA's **Viking program** in the 1970s, which consisted of two orbiters and two landers. These missions provided the first detailed images of the Martian surface and conducted experiments to search for signs of life. Although Viking did not find definitive evidence of life, its data revealed a complex landscape marked by ancient riverbeds and volcanic features, suggesting that Mars once had conditions favorable for life. The Viking missions set a precedent for subsequent exploration efforts, emphasizing the importance of robotic technology in planetary science.

The advent of rovers has revolutionized our ability to explore Mars. The Mars Exploration Rovers, Spirit and Opportunity, launched in 2003, were designed to traverse the Martian terrain, conducting extensive geological analyses. Their successful operation far exceeded expectations, with Opportunity working for nearly 15 years instead of the planned 90 days. These rovers provided compelling evidence of past water activity, including the discovery of hematite, a mineral often formed in wet environments. The rovers' findings have been crucial in shaping our understanding of Mars' climatic history and its potential habitability.

In recent years, missions such as the Curiosity rover and the Perseverance rover have continued to push the boundaries of robotic exploration. Curiosity, which landed in 2012, has been exploring **Gale Crater**, analyzing rock samples and studying the planet's climate and geology. Perseverance, which landed in 2021, is tasked with searching for signs of ancient microbial life and collecting samples for future return to Earth. This mission also includes the **Ingenuity helicopter**, which has successfully demonstrated powered flight in Mars' thin atmosphere, opening new avenues for exploration and reconnaissance.

The data collected from these robotic missions has improved our understanding of Mars and enlightened preparations for future human exploration. By identifying potential resources, such as water ice, and assessing environmental conditions, robotic explorers play a critical role in making interplanetary travel feasible. As we stand on the brink of sending humans to Mars, the insights gained from robotic exploration will be indispensable. The quest for knowledge about our neighboring planet continues, with each mission bringing us closer to understanding the mysteries of Mars and the broader implications for life beyond Earth.

Human Colonization Prospects

Human colonization of celestial bodies within the solar system has transitioned from the realm of science fiction to a plausible objective. The ongoing advancements in aerospace technology, coupled with a growing understanding of extraterrestrial environments, have opened new avenues for exploration and potential habitation. Mars, with its similarities to Earth, is often at the forefront of colonization discussions, yet other celestial bodies like the Moon, Europa, and even asteroids are increasingly considered viable candidates for human settlement.

Mars stands out due to its relatively hospitable conditions compared to other planets. With a day length similar to Earth's and the presence of polar ice caps, it offers potential resources for water and oxygen. Plans for human missions to Mars are already in development, with organizations like NASA and private companies envisioning manned missions as early as the 2030s. The establishment of a sustainable colony on Mars would require innovative solutions for food production, habitat construction, and life support systems, leveraging in-situ resource utilization to minimize reliance on Earth.

The Moon serves as another significant target for colonization. Its proximity to Earth makes it an ideal testing ground for technologies needed for longer interplanetary travel. Establishing a lunar base could facilitate research and development in a low-gravity environment, helping to prepare for future missions to Mars and beyond. The Moon's **regolith** could potentially be used for construction and as a resource for fuel, making it a strategic outpost in the quest for deeper space exploration.

Beyond Mars and the Moon, the exploration of moons like Europa, with its subsurface ocean, presents tantalizing prospects. Europa's icy crust may harbor conditions suitable for life, making it an exciting target for both scientific research and potential colonization. Establishing a human presence on Europa would provide unique challenges, such as dealing with high radiation levels and the difficulty of accessing its ocean, but the rewards could vastly expand our understanding of life beyond Earth.

As colonization efforts advance, ethical and logistical considerations will play an important role. The potential impact on extraterrestrial ecosystems, the governance of off-world colonies, and the sustainability of human life in these environments must all be carefully examined. Collaboration between nations, private entities, and scientific communities will be essential in ensuring that humanity's expansion into the solar system is conducted responsibly and effectively. The dream of colonizing other worlds is much more than a vision of the future; it is a call to action for explorers seeking to push the boundaries of human presence in the cosmos.

Composition and Types of Asteroids

Asteroids are fascinating celestial bodies that inhabit the vast expanse of our solar system, primarily found in the asteroid belt between Mars and Jupiter. These remnants from the early solar system offer valuable insights into the formation and evolution of planetary bodies. Composed mainly of rock, metal, and various ices, asteroids vary significantly in their composition, size, and structure, making them essential targets for exploration. Ongoing research into the composition and types of asteroids can enhance our knowledge of planetary science and the potential resources they may offer for future interplanetary endeavors.

The composition of asteroids can be broadly categorized into three main types: *C-type*, *S-type*, and *M-type*. **C-type**, or carbonaceous asteroids, are the most common, accounting for around **75% of known asteroids**. They are rich in carbon and other volatile compounds, resembling the primordial material from which the solar system formed. **S-type**, or silicate asteroids, are primarily composed of silicate minerals and metals, such as nickel and iron. These asteroids are found closer to the inner solar system and are often associated with the remnants of differentiated bodies that experienced geological processes. Lastly, **M-type** asteroids are metallic and consist mainly of iron and nickel, representing the cores of larger bodies that were shattered by collisions.

Asteroids also exhibit a wide variety of shapes and sizes, ranging from small boulders a few meters across to massive bodies hundreds of kilometers in diameter. The largest known asteroid, **Ceres**, is classified as a dwarf planet due to its size and spherical shape. Other notable asteroids include **Vesta**, which has a differentiated structure with a rocky crust and a metal-rich core, and **Eros**, a small, elongated asteroid that has been the subject of extensive study by space missions. The diversity in size and shape reflects the complex history of these bodies, influenced by factors such as collisions, gravitational interactions, and thermal processes.

The study of asteroids has become a central part of understanding the solar system's history and holds great promise for future exploration and resource utilization. Many asteroids contain valuable materials, including metals and water ice, which could support long-duration missions and even serve as fuel for spacecraft. The concept of [asteroid mining](#) has gained traction, as extracting resources from these bodies could alleviate the need to launch everything from Earth, making space exploration more sustainable and economically viable.

As explorers venture into the depths of the solar system, asteroids present unique opportunities for scientific discovery and technological advancement. Missions focused on asteroids, such as NASA's **OSIRIS-REx** and Japan's **Hayabusa2**, have already returned significant data and samples, revealing the complex nature of these ancient bodies. Cataloging the composition and types of asteroids serves in part to enrich our knowledge of the solar system and prepare us for the challenges and possibilities that lie ahead in our quest to explore beyond the blue planet.

Mining Opportunities

The Solar System presents a diverse array of mining opportunities, each with unique resources and challenges. As explorers venture beyond Earth, the potential for extracting valuable materials from asteroids, moons, and planets becomes increasingly apparent. Asteroid mining, in particular, has gained attention due to the abundance of precious metals and rare minerals found in these celestial bodies. With estimates suggesting that a single asteroid could contain more platinum than has ever been mined on Earth, the prospect of tapping into these resources could revolutionize industry and technology.

The Moon, our closest celestial neighbor, also offers significant mining prospects. Its surface is rich in helium-3, a potential fuel for future fusion reactors, which could provide a nearly limitless energy source. Additionally, the lunar regolith contains a variety of useful elements, including titanium and rare earth metals. The establishment of mining operations on the Moon would not only facilitate resource extraction but could also support sustained human presence and exploration, acting as a launching point for further missions into deeper space.

Mars, with its geological diversity, holds promise for mining endeavors as well. The planet's surface is rich in iron, nickel, and cobalt, which are essential for construction and manufacturing in space. Water ice deposits at the poles and beneath the surface could be harvested for life support and fuel production, making Mars a key player in the broader strategy of interplanetary colonization. The combination of in-situ resource utilization and mining could enable explorers to establish a self-sustaining presence on the Red Planet.

Beyond the inner planets, the gas giants and their moons present intriguing possibilities for mining. For instance, Jupiter's moon Europa is believed to harbor a subsurface ocean beneath its icy crust, which may contain the building blocks of life and essential resources. Saturn's moon Enceladus, known for its geysers shooting water vapor into space, could also provide access to water and organic compounds. These celestial bodies may offer scientific insights as well as valuable materials for future missions, fostering a deeper understanding of the Solar System while supporting human exploration.

As technology advances, the feasibility of mining in space becomes more attainable. Robotic mining operations, autonomous systems, and advanced propulsion methods will play crucial roles in the development of this industry. Collaborations between governmental space agencies and private enterprises are essential to create a sustainable framework for resource extraction beyond Earth. By investing in these mining opportunities, explorers can pave the way for a new era of interplanetary travel and exploration, transforming our understanding of the Solar System and expanding human presence within the Milky Way Galaxy.

Exploration Missions

The exploration missions of our solar system have evolved dramatically since the early days of space travel. From the first human footprints on the Moon to the robotic rovers traversing the Martian landscape, these missions have significantly expanded our understanding of planetary science and the potential for life beyond Earth. Each mission serves as a stepping stone, laying the groundwork for future exploration and providing invaluable data that informs our next steps into the cosmos. The quest to explore the solar system is driven by the desire to learn more about our celestial neighbors and to prepare for potential interplanetary travel.

One of the most iconic exploration missions is NASA's Apollo program, which successfully landed humans on the Moon between 1969 and 1972. The Apollo missions not only demonstrated the feasibility of human space travel but also yielded a wealth of scientific information about the Moon's geology and history. The samples collected from the lunar surface continue to provide insights into the formation of the Earth-Moon system and the broader solar system. The legacy of Apollo has inspired subsequent missions, including the Artemis program, which aims to return humans to the Moon and establish a sustainable presence there, paving the way for future missions to Mars and beyond.

Mars has become a focal point for exploration missions due to its potential to harbor life and its similarities to Earth. Robotic missions, such as the Mars rovers Spirit, Opportunity, Curiosity, and Perseverance, have provided unprecedented access to the Martian surface. These rovers have conducted extensive geological surveys, analyzed soil samples, and searched for signs of past life. The data collected from these missions is crucial for understanding Mars' climate and geology, as well as for assessing its habitability. The ongoing exploration of Mars not only enhances our knowledge of the planet but also serves as a testing ground for technologies and strategies that will be essential for human exploration of the Red Planet.

Beyond Mars, the exploration of outer planets and their moons has revealed a diverse and captivating solar system. Missions like Voyager, **Galileo**, and **Cassini** have provided stunning images and data about Jupiter, Saturn, and their intricate systems of moons. Europa, one of Jupiter's moons, has garnered particular interest due to its subsurface ocean, which may harbor the conditions necessary for life. Similarly, Saturn's moon Enceladus has shown signs of geysers erupting from its icy surface, suggesting a dynamic environment beneath. These discoveries continue to illustrate the complexity of our solar system and highlight the potential for future missions to explore these intriguing worlds in search of extraterrestrial life.

As we look to the future of exploration missions, the prospect of interplanetary travel becomes increasingly feasible. Advances in propulsion technology, life support systems, and robotics are paving the way for missions that were once thought to be science fiction. Initiatives such as NASA's Artemis program and private ventures like SpaceX's plans for Mars colonization aim to establish a human presence beyond Earth. Additionally, international collaborations, such as the Lunar Gateway and the **Mars Sample Return mission**, are essential in pooling resources and expertise for future exploration. The next era of exploration missions promises to deepen our understanding of the solar system and potentially unlock the secrets of our galaxy, bringing humanity one step closer to becoming an interstellar species.

Jupiter's Atmosphere and Weather Systems

From NASA we learn that: "".

Jupiter's atmosphere is a marvel of the solar system, characterized by its intricate layers and dynamic weather systems. Composed primarily of hydrogen and helium, with traces of methane, ammonia, hydrogen sulfide, and water, the atmosphere exhibits a complex structure that is both fascinating and daunting. The uppermost layer features clouds of ammonia ice, while deeper layers transition into thicker clouds of water vapor, creating a rich palette of colors that can be seen from telescopes on Earth. This striking appearance is largely due to the varying compositions and temperatures at different altitudes, leading to the formation of bands, storms, and vortices that define Jupiter's meteorological phenomena.

The most recognizable feature of Jupiter's atmosphere is the Great Red Spot, a massive storm that has been raging for at least **350 years**. This **anticyclonic storm** is larger than Earth, with wind speeds that can reach up to **432 kilometers per hour**. The Great Red Spot is a prime example of the extreme weather systems that develop in Jupiter's atmosphere, driven by its rapid rotation and immense energy output. The planet completes a rotation approximately every 10 hours, which contributes to the formation of powerful jet streams that create the distinct banding of its atmosphere. These jet streams are responsible for the continuous movement of clouds and storms, making Jupiter's weather systems both dynamic and unpredictable.

In addition to the Great Red Spot, Jupiter's atmosphere features an array of smaller storms and weather patterns, many of which are short-lived but intense. These include white ovals, brown barges, and other storm systems that exhibit unique characteristics. The interaction between these storms and the planet's magnetic field further complicates the atmospheric dynamics. The magnetic field generates auroras at the poles, with charged particles colliding with the atmosphere to create spectacular light displays. These phenomena while being visually captivating also provide insight into the complex processes occurring within Jupiter's atmosphere.

The study of Jupiter's atmosphere is not just important for understanding the gas giant itself but also the formation and evolution of planetary atmospheres as a whole. Space missions, such as NASA's **Juno spacecraft**, have provided unprecedented data on the planet's gravitational field, magnetic field, and atmospheric dynamics. These observations reveal the depth of Jupiter's atmosphere, showing that the weather patterns extend far below the cloud tops and are influenced by factors such as internal heat and composition. Such insights enhance our knowledge of how gas giants function and may offer clues about the atmospheres of exoplanets located in distant solar systems.

Explorers venturing to Jupiter must consider the challenges posed by its extreme atmosphere and weather systems. The thick atmosphere, coupled with intense radiation belts, presents significant hazards for spacecraft and human exploration. Tracking these atmospheric conditions is crucial for planning missions that would study the planet further or explore its intriguing moons, such as Europa, **Ganymede**, and **Callisto**. As we continue to unlock the secrets of Jupiter's atmosphere, we pave the way for a deeper understanding of the solar system's largest planet and its myriad of phenomena, enriching our journey through the Milky Way Galaxy.

The Moons of Jupiter

Jupiter, the giant of our solar system, is not only remarkable for its immense size and powerful storms but also for its captivating collection of moons. With o**ver 79 known moons**, these celestial bodies vary widely in size, composition, and potential for exploration. Among them, four stand out due to their unique characteristics and proximity to the gas giant: **Io**, Europa,

Ganymede, and Callisto. Known collectively as the **Galilean moons**, they were **first observed by Galileo Galilei in 1610** and have since become focal points for scientific inquiry and exploratory missions.

Io, the innermost of the Galilean moons, is a geological wonder. It is the most volcanically active body in the solar system, with hundreds of volcanoes dotting its surface. The intense heat that drives this activity is generated by tidal forces exerted by Jupiter's immense gravity. Explorers consider the potential for interplanetary travel realize that Io's dynamic environment offers unique insights into planetary geology and the processes that shape celestial bodies. However, the harsh conditions, including radiation levels far exceeding those found on Earth, pose significant challenges for exploration.

Europa, the next moon outward, presents one of the most intriguing possibilities for harboring life. Beneath its icy crust lies a subsurface ocean, which may contain more than twice the amount of water found on Earth. The potential for this ocean to support life has spurred interest in missions aimed at drilling through the ice to explore the ocean below. Europa's surface features, including linear ridges and chaotic terrain, suggest a dynamic history, raising questions about the moon's past and the possibility of extraterrestrial ecosystems. For explorers, Europa represents a tantalizing destination in the quest to understand life beyond our planet.

Ganymede, the largest moon in the solar system, is unique for its size and geological diversity. It is the only moon known to have its own magnetic field, which may be indicative of a partially liquid iron or iron-sulfide core. Ganymede's surface is a mix of two types of terrain: bright, icy regions and darker, heavily cratered areas, suggesting a complex geological history. Its potential for hosting a subsurface ocean adds to its appeal for exploration. In planning for future missions, Ganymede presents opportunities to study both its surface and interior, offering insights into the formation of moons and planets in the solar system.

Callisto, the outermost of the Galilean moons, is often described as a "frozen world." Its heavily cratered surface indicates that it has remained relatively unchanged for billions of years, providing a window into the early solar system. Callisto's low radiation environment and potential subsurface ocean make it a compelling candidate for exploration, particularly for human missions. As explorers venture into the depths of the solar system, understanding Callisto's history and composition could yield valuable information about the evolution of celestial bodies and the conditions necessary for life. Each of these moons holds secrets waiting to be uncovered, urging explorers to journey beyond the blue planet and investigate the mysteries of the solar system.

Future Missions to Jupiter

Future missions to Jupiter promise to unlock the secrets of the largest planet in our Solar System, offering explorers new insights into its complex atmosphere, magnetic field, and diverse moons. As our understanding of space exploration evolves, several missions are in the pipeline, aiming to deepen our knowledge of this gas giant. These missions will not only focus on Jupiter itself but also on its intriguing moons, particularly Europa, Ganymede, and Callisto, which are believed to harbor conditions suitable for life.

NASA's Europa Clipper mission, slated for launch in the 2020s, exemplifies the exciting future of **Jovian** exploration. This mission will conduct detailed reconnaissance of Europa's icy surface and subsurface ocean. Equipped with a suite of scientific instruments, Europa Clipper will analyze the moon's composition, study its geologic activity, and assess its potential habitability. The data gathered will be central in understanding whether Europa could support life, drawing explorers closer to answering one of humanity's most profound questions.

The **European Space Agency**'s **JUICE** (JUpiter ICy moons Explorer) mission, also set to launch in the coming years, will focus on Ganymede, Europa, and Callisto. JUICE aims to investigate the icy crusts of these moons, their subsurface oceans, and their potential for hosting life. With advanced technology and instruments, the mission will also study Jupiter's atmosphere and **magnetosphere**, allowing researchers to piece together the intricate relationships between the planet and its moons. This dual focus enhances our understanding of celestial bodies in the outer Solar System.

In addition to these missions, the potential for collaboration between international space agencies could lead to more comprehensive explorations of Jupiter and its moons. Joint missions would allow for shared resources, expertise, and technology, which could enhance the scientific yield of future endeavors. Such collaborations could also pave the way for innovative exploration techniques, such as orbiters, landers, or even potential sample-return missions from Europa or Ganymede, significantly enriching our understanding of these distant worlds.

In planning for what lies ahead, the prospect of human missions to Jupiter may also materialize, although they remain a more distant goal. The challenges posed by the planet's intense radiation, extreme temperatures, and its vast distance from Earth are significant. However, advancements in technology and a growing interest in interplanetary travel may eventually make such missions feasible. The exploration of Jupiter and its moons represents a tantalizing scientific frontier as well as an opportunity for humanity to extend its reach into the cosmos, inspiring the next generation of explorers to dream beyond the blue planet.

Saturn's Rings and Moons

Saturn, the sixth planet from the Sun, is renowned for its spectacular rings and diverse collection of moons, making it a focal point for explorers intrigued by the dynamics of our solar system. The planet itself is a gas giant, primarily composed of hydrogen and helium, and is characterized by its striking yellowish hue. However, what truly sets Saturn apart are its prominent rings, which extend up to 175,000 miles from the planet's center and comprise countless particles made primarily of ice and rock. These visually stunning rings provide valuable insights into the processes of planetary formation and the evolution of celestial bodies.

The rings of Saturn are divided into several distinct sections, including the , each differing in thickness, density, and composition. The , the outermost, is separated from the B ring by the , a gap created by gravitational interactions with Saturn's moons. The is the widest and brightest, while the , closer to the planet, is less dense and more translucent. The structure of these rings is continually shaped by the gravitational pull of Saturn's moons, demonstrating a delicate balance between attraction and centrifugal force. Exploring these rings offers an opportunity to understand the intricate mechanics of celestial systems and the forces that govern them.

Saturn is home to , each with unique characteristics and potential for exploration. , the largest of Saturn's moons, is particularly intriguing due to its dense atmosphere and the presence of liquid methane lakes on its surface. Titan's environment bears similarities to early Earth, making it a prime candidate for studying the prebiotic conditions that could lead to life. Other significant moons include Enceladus, which has geysers that eject water vapor and organic molecules, suggesting a subsurface ocean that may harbor the conditions for life. The diverse geological features of these moons provide a treasure trove of information for explorers seeking to understand the possibilities of life beyond our planet.

The interactions between Saturn's rings and its moons create a dynamic environment that is both complex and fascinating. As moons pass through the rings, they can induce waves and structures within the ring material, showcasing a lively interplay of gravitational forces. This interaction not only improves our understanding of ring dynamics and highlights the importance of moons in shaping their planetary environments. The ongoing exploration of these phenomena through spacecraft missions, such as the , has provided unparalleled insights into the complexities of Saturn's ring system and its moons.

For those embarking on interplanetary travel and exploration, Saturn and its rings offer a captivating destination that is rich in scientific inquiry and discovery. The combination of the majestic rings and the intriguing moons presents not only aesthetic wonder but also profound questions about the origins of our solar system and the potential for life beyond Earth. Explorers who dare to venture beyond the blue planet have come to know that Saturn stands as a beacon of knowledge and mystery, enticing them to unlock the secrets of the cosmos.

The Cassini-Huygens Mission

The Cassini-Huygens mission stands as one of the most ambitious and successful endeavors in the history of space exploration. Launched in 1997, this collaborative project between NASA, the European Space Agency (ESA), and the was designed to study Saturn, its rings, and its numerous moons. The mission comprised two primary components: the , which would remain in orbit around Saturn, and the , intended to descend through Titan's thick atmosphere and land on its surface. This unprecedented dual approach allowed scientists to gather a wealth of data about one of the most intriguing systems in our solar system.

Cassini entered orbit around Saturn in July 2004, and from that vantage point, it conducted a detailed analysis of the planet's atmosphere, magnetic field, and gravitational field. One of the primary objectives was to investigate the complex structure and composition of Saturn's rings. Cassini's detailed observations revealed intricacies such as the presence of tiny moons within the rings, the dynamics of ring particles, and the processes that shape their appearance. The orbiter sent back stunning images and data that enhanced our understanding of planetary ring systems, providing insights that extended beyond Saturn to inform theories about other celestial bodies.

In January 2005, the Huygens probe made history by successfully landing on Titan, Saturn's largest moon. This marked the first landing on a body in the outer solar system. Huygens descended through Titan's dense atmosphere, collecting data on atmospheric composition and conditions. Upon landing, it transmitted information about the moon's surface, revealing a landscape marked by rivers and lakes of liquid methane and ethane. These findings suggested that Titan, with its complex chemistry and geological processes, could offer clues about prebiotic conditions, making it a target of interest for astrobiology.

The Cassini-Huygens mission also made significant discoveries regarding Saturn's moons. Enceladus, in particular, captured the attention of scientists due to its geysers that eject water vapor and ice particles into space. Cassini's observations indicated that beneath its icy crust, Enceladus harbors a subsurface ocean, raising the possibility of conditions suitable for life. Other moons, such as and , were also studied, revealing a diverse array of geological features and histories that challenge our understanding of moon formation and evolution in the solar system.

The mission concluded in September 2017, but its legacy continues to influence ongoing and future exploration. The data collected over two decades has paved the way for new hypotheses regarding planetary formation and evolution, not just within our solar system but also in the broader context of exoplanet studies. As explorers look to the future, the Cassini-Huygens mission serves as a reminder of the importance of collaborative efforts in space exploration and the profound discoveries that await in the vastness of the Milky Way Galaxy.

Potential for Future Exploration

The potential for future exploration within our solar system is vast and filled with opportunities that could redefine our understanding of planetary science and interstellar possibilities. As technology advances, missions to diverse celestial bodies are becoming increasingly feasible. These missions could provide invaluable data on the composition, atmosphere, and potential for life on planets and moons previously thought unreachable. With instruments becoming more sensitive and spacecraft more robust, the exploration of destinations like Europa, Enceladus, and even the gas giants themselves is on the horizon, promising to unveil secrets hidden beneath icy crusts and thick atmospheres.

Mars remains at the forefront of future exploration endeavors. As various nations and private entities invest in technology to send humans to the Red Planet, the potential for long-term habitation, resource extraction, and scientific discovery grows. Missions aimed at studying Martian geology, climate, and potential biosignatures will not only inform us about Mars' past but also prepare us for the challenges of human settlement. The establishment of bases on Mars could serve as launching points for deeper exploration into the solar system, creating a new era of interplanetary travel that connects Earth with our neighboring worlds.

The outer solar system presents another frontier for exploration. The gas giants, such as Jupiter and Saturn, harbor numerous moons that are among the most intriguing targets for scientific investigation. These moons may possess subsurface oceans, which could potentially host microbial life. Future missions to these distant worlds will require innovative propulsion technologies and advanced robotics to endure the harsh environments. The study of these moons can offer insight into planetary formation and the potential for life beyond Earth, igniting the imaginations of explorers eager to uncover the mysteries of the universe.

The concept of interstellar travel is gradually shifting from science fiction to a more tangible goal as researchers explore theoretical technologies such as and . While the challenges are immense, the pursuit of such technologies could one day enable humanity to venture beyond our solar system and explore the Milky Way Galaxy. This endeavor promises not only to expand our reach but also to answer fundamental questions about our place in the universe. Such exploration could lead to encounters with exoplanets that may bear similarities to Earth, further stimulating the search for extraterrestrial life.

Regarding plans for future exploration, collaboration among nations and private enterprises will be essential for the success of these ambitious endeavors. Sharing knowledge, resources, and technologies can accelerate the pace of discovery and broaden the scope of exploration. With a renewed sense of purpose and a unified vision, the next generation of explorers is poised to embark on a journey that transcends the boundaries of our planet, pushing the limits of what is known and venturing boldly into the uncharted territories of the solar system and beyond. The potential for future exploration is not just a promise; it is a call to action for all who dare to dream of what lies beyond the blue planet.

Chapter 10: Uranus and Neptune: The Ice Giants

Characteristics of Ice Giants

Ice giants, a classification that includes Uranus and Neptune, possess distinct characteristics that set them apart from their gas giant counterparts. These planets are primarily composed of water, ammonia, and methane in various states, giving them unique atmospheric and surface features. Unlike the more massive gas giants, which have thick hydrogen and helium envelopes, ice giants exhibit a greater proportion of heavier elements, influencing their internal structure and thermal dynamics.

The atmospheres of ice giants are characterized by dynamic weather patterns and striking cloud formations. Uranus, for instance, displays a relatively bland atmosphere with faint clouds, while Neptune showcases vibrant blue hues due to the absorption of red light by methane. Both planets experience intense winds, with Neptune boasting the fastest winds recorded in the solar system, reaching speeds of up to 1,500 kilometers per hour. These atmospheric conditions contribute to their unique climatic features and offer insights into planetary weather systems.

In terms of internal structure, ice giants have a differentiated composition that includes a rocky core surrounded by a mantle of icy materials. This configuration leads to unique magnetic field characteristics, which are tilted significantly relative to their rotational axes. The magnetic fields of Uranus and Neptune do not align with their geometric centers, suggesting complex internal dynamo processes. The interaction between their magnetic fields and solar wind creates auroras and contributes to the planets' overall magnetosphere dynamics.

The exploration of ice giants is of particular interest to planetary scientists and explorers. Their unique compositions and atmospheric phenomena provide clues about the formation and evolution of the solar system. Missions to these distant worlds could bring to the surface the mysteries of their internal structures, atmospheres, and magnetic fields. Current technological advancements in space exploration pave the way for future missions that could provide valuable data and enhance our understanding of these enigmatic planets.

In addition to their scientific significance, ice giants hold potential for future exploration and resource utilization. The presence of water, ammonia, and other compounds may offer opportunities for in-situ resource utilization, which may support long-term human exploration in the outer solar system. Continued research into the characteristics of ice giants is essential for explorers seeking to uncover the secrets of our cosmic neighborhood and establish a foothold beyond the terrestrial realm.

The Mysteries of Uranus

The planet Uranus, the seventh from the Sun, presents a unique set of mysteries that tantalize astronomers and explorers alike. Unlike its more vibrant counterparts, Uranus is often described as a pale blue or greenish sphere, a result of the methane in its atmosphere that absorbs red light. This distant ice giant is not only intriguing due to its color but also because of its unusual axial tilt of approximately 98 degrees. This extreme tilt causes Uranus to rotate on its side, leading to extreme seasonal variations that have fascinated scientists since the planet was discovered in 1781 by **Sir William Herschel**.

One of the most compelling mysteries of Uranus lies in its internal structure and atmospheric dynamics. Unlike gas giants such as Jupiter and Saturn, Uranus is classified as an ice giant, characterized by its icy composition of water, ammonia, and methane. This distinction raises questions about the formation and evolution of the planet. The lack of a well-defined solid surface and the presence of a thick atmosphere contribute to the challenges of studying its interior.

Researchers are particularly interested in understanding the planet's core, which is believed to be composed of a rocky and icy mixture, potentially offering clues about the formation of other celestial bodies in the solar system.

Uranus's complex weather patterns also add to its allure. The planet exhibits high-speed winds that can reach up to 560 miles per hour, making it one of the windiest places in the solar system. These winds are especially notable during its summer months, when storms and atmospheric phenomena become more pronounced. The planet's unique tilt results in extreme seasonal changes, leading to periods of intense activity and calm. Ongoing research into these atmospheric conditions continue to enhance our understanding of Uranus and provide insight into the dynamics of other gas and ice giants in the Milky Way.

The exploration of Uranus has been limited, with only one spacecraft, Voyager 2, having flown by the planet in 1986. This historic encounter provided invaluable data and images, revealing Uranus's faint rings and numerous moons. However, much remains unknown. The planet's moons, such as **Titania** and **Oberon**, hold their own mysteries, with some displaying geological features that suggest a past history of tectonic activity. Future missions to Uranus and its moons could unlock secrets about the planet's formation, its atmospheric processes, and the potential for understanding other similar worlds in the galaxy.

With the hope of future explorations emerging, the mysteries of Uranus present an exciting opportunity for deeper investigation. The prospect of sending an orbiter or lander to study the planet and its moons in detail could revolutionize our understanding of ice giants. Such missions may eventually answer lingering questions about Uranus and provide broader insights into the origins of the solar system and the potential for life in other planetary systems. Moving forward on our path through the cosmos, the mystifying Uranus stands as a beacon for exploration, urging us to peer deeper into the unknown.

Exploring Neptune and Its Moons

Neptune, the eighth planet from the Sun, is a distant and cryptic giant located about 30 **astronomical units** from Earth. With its striking blue hue, a result of methane in its atmosphere, Neptune is often referred to as the "blue planet" of the outer solar system. It is the fourth largest planet by diameter and has a mass that is **17 times greater than Earth's**. Despite its distance and the challenges that come with exploring such a remote world, Neptune presents a fascinating opportunity for interplanetary explorers seeking to understand the dynamics of our solar system.

The atmosphere of Neptune is dynamic and turbulent, characterized by the fastest winds recorded in the solar system, reaching speeds of up to 1,500 miles per hour. This extreme weather is driven by the planet's internal heat and contributes to its complex weather patterns, including large storms reminiscent of Jupiter's Great Red Spot. The most famous storm on Neptune, known as the **Great Dark Spot**, was observed by the Voyager 2 spacecraft in 1989, showcasing the planet's active climate. Collecting more data on these atmospheric conditions could reveal insights into the processes that govern not only Neptune but also other gas giants.

Neptune's system of moons adds another layer of intrigue to this distant planet. There are **14 known moons**, with **Triton** being the largest and most remarkable. Triton is unique because it is the only large moon in the solar system that orbits its planet in a **retrograde direction**, suggesting that it may have been captured by Neptune's gravity rather than forming in place. This unusual orbit adds to the mystery surrounding Triton's origin and its geological history. The surface of Triton features geysers that erupt nitrogen gas, indicating possible geological activity beneath its icy crust, making it a prime candidate for future exploration.

Another noteworthy moon is **Proteus**, which, despite being irregularly shaped and heavily cratered, provides valuable insights into the evolutionary history of the Neptunian system. Smaller moons like **Nereid** and **Larissa** also contribute to our understanding of the diverse and complex interactions within Neptune's gravitational field. Each moon presents unique characteristics that could inform scientists about the formation of the outer solar system and the conditions that led to the development of different celestial bodies.

As our capabilities in space exploration continue to advance, the potential for missions to Neptune and its moons becomes increasingly feasible. Future explorers may utilize advanced robotic spacecraft or even crewed missions to study these distant worlds up close. The knowledge gained from such missions should enhance our understanding of Neptune and provide critical information about the broader dynamics of icy bodies in the outer solar system. With each passing year, the dream of exploring Neptune and its captivating moons draws closer, promising to unveil the mysteries of one of the solar system's most intriguing destinations.

Understanding Our Galaxy

The Milky Way Galaxy, our cosmic home, is a vast and intricate structure that stretches approximately **100,000 light-years in diameter**. It is a **barred spiral galaxy**, characterized by its rotating disk, spiral arms, and central bulge, which houses a **supermassive black hole** known as **Sagittarius A*** (Star). This central region is densely packed with stars, gas, and dust, providing a rich environment for astrophysical research and exploration. Each component of the galaxy plays a vital role in its overall dynamics and evolution, making it essential for explorers to understand the structure and behavior of this celestial entity.

At the heart of the Milky Way lies the **galactic core**, where intense gravitational forces govern the motion of surrounding stars and other matter. The core is surrounded by several dense stellar clusters, while the spiral arms, which contain a higher concentration of stars and interstellar material, emerge from the bulge. These arms are the sites of star formation, where clouds of gas and dust collapse under their own gravity to create new stars. The processes that occur in these regions can provide insights into the life cycle of stars and the evolution of planetary systems, including our own.

Our galaxy is not just a collection of stars; it also contains a variety of celestial phenomena that captivate the imagination of explorers. Among these are **nebulae**, which are colorful clouds of gas and dust that often serve as nurseries for new stars. Additionally, there are various types of **star clusters**, including **globular clusters**, which are tightly packed groups of ancient stars, and **open clusters**, which are more loosely bound and younger. Each of these features offers a unique opportunity for exploration and study, revealing the diverse environments that exist within our galaxy.

Another intriguing aspect of the Milky Way is its interaction with neighboring galaxies. The **Andromeda Galaxy**, for instance, is on a collision course with our own, expected to merge in about 4.5 billion years. This cosmic dance highlights the dynamic nature of galaxies and their constituents. Being able to track and predict these interactions can be eye-opening for explorers interested in the future of our galaxy and the potential for discovering new worlds formed as a result of such mergers. These events can lead to the redistribution of stars and gas, potentially triggering new waves of star formation.

On our adventures beyond this blue planet and into the depths of the Milky Way, the importance of understanding our galaxy becomes increasingly clear. The knowledge gained from studying its structure, stellar populations, and interactions with other galaxies continually enriches our understanding of the universe and enhances our ability to explore it. Armed with this insight, explorers can embark on journeys that extend the boundaries of human knowledge and experience, bringing to light the secrets held within our galactic home and paving the way for future interplanetary travel and exploration.

Potential Exoplanets

The search for potential **exoplanets** has long captivated astronomers and explorers, as these distant worlds hold the promise of new discoveries and possibly life beyond our own planet. An exoplanet, or **extrasolar planet**, is any planet that orbits a star outside our solar system. Advances in technology and observational techniques have allowed scientists to identify thousands of these celestial bodies, expanding our understanding of the universe and opening up new avenues for interplanetary travel and exploration.

One of the most exciting aspects of exoplanet research is the diverse range of environments these planets may possess. For instance, some exoplanets lie within the **habitable zone** of their stars, where conditions may be just right for liquid water to exist—a key ingredient for life as we know it. With the launch of space telescopes such as **Kepler** and **TESS**, the hunt for Earth-like planets has intensified, revealing a variety of planetary systems that challenge our existing theories of planetary formation and habitability.

Among the notable findings is the **TRAPPIST-1 system**, which contains seven Earth-sized planets, three of which are located in the habitable zone. This discovery has sparked interest in the potential for future missions to explore these worlds. The prospect of landing on an exoplanet and studying its atmosphere, geology, and potential biosignatures raises thrilling possibilities for explorers. As technology continues to evolve, the dream of sending probes or even humans to these distant destinations becomes increasingly within reach.

Another intriguing category of exoplanets is the "**hot Jupiters**," gas giants that orbit very close to their stars. While these immense worlds may not be suitable for human exploration or habitation, their study provides insights into planetary formation and migration. Understanding how these massive planets develop and interact with their stellar environments can inform our knowledge of system dynamics, which is vital for planning future interstellar missions.

The exploration of potential exoplanets has been building up our scientific knowledge and also speaks to the innate human desire to explore the unknown. As we refine our exploration techniques and expand our capabilities, the Milky Way Galaxy holds countless mysteries waiting to be disclosed. With each new discovery, we inch closer to answering profound questions about our place in the cosmos and the potential for life beyond Earth, making the pursuit of exoplanets a cornerstone of future interplanetary exploration.

The Future of Galactic Exploration

The future of galactic exploration is poised to transform our understanding of the universe and our place within it. As technology advances, the dream of interplanetary travel evolves from science fiction to a tangible reality. Initiatives from both governmental space agencies and private enterprises are leading the way for missions beyond our home planet. With the successful deployment of more sophisticated telescopes and spacecraft, the next few decades promise to broaden our horizons, enabling explorers to venture further into the cosmos than ever before.

One of the most significant advancements in galactic exploration will be the development of [improved propulsion technologies](). Traditional chemical rockets, while effective for current missions, are limited by speed and fuel efficiency. Researchers are exploring alternatives such as ion propulsion, **nuclear thermal engines**, and even theoretical concepts like the **Alcubierre warp drive**. These innovations aim to reduce travel time within our solar system, making destinations like Mars and the moons of Jupiter and Saturn more accessible. As these technologies mature, the prospect of human settlement on other planets becomes increasingly plausible.

Interplanetary travel will also be influenced by advancements in robotics and artificial intelligence. Autonomous rovers and drones can explore hostile environments, collect data, and conduct scientific experiments without the need for constant human oversight. These robotic explorers will serve as precursors to human missions, scouting terrain and identifying resources. They will play a crucial role in preparing for human habitation by mapping out potential sites for bases and gathering information on local conditions. As we refine these technologies, the synergy between human and robotic exploration will enhance our capabilities and ensure safer missions.

Collaboration among nations and private entities is essential for the future of galactic exploration. The **International Space Station** has already demonstrated the benefits of cooperative efforts in space, and similar partnerships will be vital as we push further into the solar system. Shared resources, knowledge, and expertise can accelerate the development of new technologies and reduce costs. Moreover, international agreements governing the exploration and use of celestial bodies will be necessary to prevent conflicts and ensure that space exploration remains a pursuit for all of humanity.

The implications of successful galactic exploration extend beyond scientific knowledge; they challenge our philosophical perspectives and inspire future generations. As humans explore new worlds, they will confront questions about life, existence, and the ethical considerations of colonization. The exploration of the Milky Way galaxy may one day lead to the discovery of extraterrestrial life, fundamentally altering our understanding of biology and our place in the universe. By promoting a spirit of curiosity and cooperation, the future of galactic exploration can ignite a passion for discovery that resonates across humanity, encouraging explorers to reach for the stars.

Propulsion Systems

Propulsion systems are fundamental to interplanetary travel and exploration within our solar system. These systems enable spacecraft to escape Earth's gravitational pull, navigate through the vacuum of space, and ultimately reach distant celestial bodies. Familiarity with various propulsion technologies can help to spark the necessary innovations for explorers who wish to venture beyond our home planet. Current propulsion methods are primarily categorized into , , and propulsion systems, each offering unique advantages and limitations in terms of efficiency, speed, and mission suitability.

Chemical propulsion remains the most widely used technology for space travel today. It relies on the combustion of propellants to produce thrust, propelling spacecraft into orbit and beyond. The classic rocket engines, such as those used in the **Saturn V** and **Space Shuttle**, exemplify this method. While chemical propulsion provides significant thrust, enabling rapid launches, it has limitations in fuel efficiency. Consequently, missions that require long-duration travel or extensive maneuvering often seek alternative propulsion systems that can provide sustained thrust over extended periods.

Electric propulsion systems are gaining prominence due to their high efficiency and ability to operate for longer durations. These systems, including ion thrusters and **Hall effect thrusters**, utilize electric energy to accelerate ions, generating thrust. This technology allows spacecraft to gradually build up speed over time, making it suitable for missions to the outer planets where fuel constraints are critical. Notable missions like NASA's **Dawn spacecraft** have successfully demonstrated the capabilities of electric propulsion, allowing for extensive exploration of celestial bodies while minimizing fuel usage.

Advanced propulsion systems, such as nuclear thermal and **solar sails**, present exciting possibilities for future exploration. Nuclear thermal propulsion uses a nuclear reactor to heat a propellant, offering higher efficiency and greater thrust than traditional chemical rockets. This technology has the potential to significantly reduce travel time to distant destinations, such as Mars or the outer planets. Solar sails, on the other hand, harness the pressure of sunlight for propulsion. By deploying large, reflective sails, spacecraft can gradually accelerate using the force of photons. This method is particularly appealing for long-term missions, as it requires no onboard fuel, allowing explorers to travel vast distances with minimal resources.

In preparation for our journeys into the vastness of the solar system and beyond, the continuous development and refinement of propulsion systems will play a crucial role in shaping the future of space travel. Mastering the intricacies of these technologies will empower explorers to select the most suitable options for their missions, ensuring a safe and efficient journey through the cosmos. The quest for advanced propulsion methods will almost certainly boost our ability to explore the Milky Way and enlarge our understanding of the universe and our place within it.

Spacecraft Design

Spacecraft design is a predominant aspect of interplanetary travel and exploration, serving as the bridge between Earth and the vast unknowns of the Solar System. Each spacecraft must be meticulously engineered to withstand the harsh environments of space, including extreme temperatures, radiation, and **microgravity**. The materials used in construction must not only be lightweight to conserve fuel but also strong enough to endure the stresses of launch and the potential impacts of **micrometeoroids**. Innovative design techniques, such as **modular construction**, allow for easier repairs and upgrades, which are essential for long-duration missions.

One of the most important considerations in spacecraft design is propulsion. Various propulsion systems, such as chemical rockets, ion drives, and solar sails, are utilized based on the mission goals and the distance to be covered. Chemical rockets provide the thrust needed for launch and initial escape from Earth's gravity, while ion drives offer fuel efficiency for long-haul space travel, allowing spacecraft to accelerate gradually over extended periods. Solar sails harness sunlight for propulsion, presenting a promising alternative for future deep-space missions where conventional fuel may be impractical.

Thermal control systems are another vital component of spacecraft design. Spacecraft experience extreme temperature fluctuations as they travel between the inner and outer Solar System. To manage these conditions, engineers incorporate insulation materials, radiators, and heaters to maintain optimal operating temperatures for sensitive equipment and crew habitats. Proper thermal regulation helps to safeguard the integrity of the spacecraft and ensures that scientific instruments can function effectively while gathering crucial data about other celestial bodies.

The design of a spacecraft's systems for life support and habitation is particularly important for missions involving human crews. These systems must provide a stable atmosphere, recycle water, and manage waste, all while ensuring the health and safety of the astronauts. Advanced life support systems use closed-loop technologies to maximize resource efficiency, minimizing the amount of supplies that need to be sent from Earth. In addition, the design of living quarters and workspaces must prioritize comfort and psychological well-being, as crew members may spend extended periods in confined environments far from home.

The inclusion of advanced technology such as autonomous navigation and artificial intelligence is revolutionizing spacecraft design. These systems enable spacecraft to make real-time decisions, navigate through complex environments, and conduct scientific experiments without constant human oversight. With explorers venturing further into the Milky Way Galaxy, the ability to operate independently will be decisive for the success of missions to distant planets and moons. The ongoing evolution of spacecraft design will pave the way for a new era of exploration, bringing the wonders of our Solar System within reach.

Life Support Systems

Life support systems are mandatory components for successful interplanetary travel, as they ensure the survival of explorers in the harsh environments of space and on other celestial bodies. These systems are designed to provide essential resources such as oxygen, water, food, and temperature regulation while removing carbon dioxide and other waste products. In the context of missions beyond Earth, understanding and optimizing these systems is paramount to prolonging human presence in space, whether on a spacecraft, space station, or planetary surface.

At the core of life support systems is the management of atmospheric conditions. In space, the absence of a breathable atmosphere means explorers rely on technology to supply oxygen. Systems like the **Oxygen Generation System** (OGS) convert water into oxygen through **electrolysis**, ensuring a continuous supply for crew members. Additionally, **scrubbers** remove carbon dioxide from the air, maintaining a safe and breathable environment. The challenge lies in creating systems that are both efficient and sustainable, minimizing reliance on resupply missions from Earth.

Water reclamation is another indispensable aspect of life support. Given the logistical challenges and costs associated with transporting water across vast distances, recycling systems become essential. Advanced filtration and purification technologies can reclaim water from various sources, including moisture in the air and even urine. For long-duration missions, such as those planned for Mars, these systems must operate with high reliability and low maintenance to support the health and hydration of explorers without frequent resupply.

Food production in space represents a frontier of its own. Traditional methods of storing and transporting food are impractical for long missions due to spoilage and bulk. Innovative solutions are being explored, including **hydroponics** and **aeroponics**, which allow for the cultivation of crops in controlled environments using minimal resources. These systems serve to provide nutrition while also contributing to psychological well-being by allowing explorers to engage in agriculture, a sense of normalcy and connection to Earth.

Finally, temperature regulation is crucial for maintaining the comfort and safety of explorers. Spacecraft and habitats must be equipped with thermal control systems that manage heat generated by equipment and the external environment. Advanced insulation materials, radiators, and active heating and cooling systems work in concert to create a stable living environment. As explorers journey deeper into the solar system, the engineering of life support systems will continue to evolve, playing a key role in enabling humanity's quest to explore the stars.

Building Habitats on Other Planets

Building habitats on other planets presents a significant challenge and opportunity for explorers of the solar system. As humanity sets its sights beyond Earth, the need for sustainable living environments on celestial bodies becomes paramount. Each planet and moon presents unique conditions that must be carefully considered. From the harsh surface of Mars to the icy crust of Europa, the design and construction of habitats will require innovative engineering, advanced technology, and a deep understanding of extraterrestrial environments.

Mars, often seen as the most viable candidate for human colonization, has conditions that both challenge and inspire. Its thin atmosphere, composed mostly of carbon dioxide, provides limited protection from cosmic radiation and requires habitats to be well-insulated and capable of recycling air and water. Explorers envision using local materials, such as Martian regolith, to create structures that can withstand temperature extremes and dust storms. Concepts such as and showcase the potential for creating safe living spaces while minimizing the need to transport materials from Earth.

In contrast, building habitats on the moons of gas giants, such as Europa and Titan, involves different considerations. Europa's subsurface ocean offers the alluring possibility of utilizing local water resources, but the icy surface poses challenges for construction and access. Habitats on Europa would need to be insulated against extreme cold and designed to withstand the intense radiation levels from Jupiter. Titan, with its dense atmosphere and lakes of liquid methane, presents an alternative environment where habitats could float on the surface, utilizing buoyancy for stability. These unique characteristics call for tailored approaches to habitat design that align with the specific conditions of each celestial body.

The role of technology in constructing these habitats cannot be overstated. Innovations in robotics, material science, and energy generation will be at the forefront. Robotic systems could be deployed to prepare sites for human arrival, conducting preliminary surveys and even assembling structures autonomously. Advanced materials, such as those developed to withstand extreme temperatures or radiation, will be essential in ensuring the longevity and safety of these habitats. Furthermore, sustainable energy solutions, such as solar panels and nuclear power, will be requisite for supporting life and operations on distant worlds.

The quest to build habitats on other planets is not merely about survival; it is about creating a new home for humanity beyond Earth. As explorers, the pursuit of knowledge and the experience of living on another planet will reshape our understanding of life and our place in the universe. The challenges of constructing these habitats will inevitably lead to advancements in science and technology, pushing the boundaries of what is possible and inspiring future generations to venture beyond the blue planet into the vast expanse of the solar system and beyond.

The Role of International Collaboration

International collaboration plays an all-important role in the exploration of the Solar System and the broader Milky Way Galaxy. As humanity looks beyond Earth, the complexities and challenges of interplanetary travel necessitate a unified approach that transcends national boundaries. Space exploration is not only a scientific endeavor but also a global initiative that requires shared resources, expertise, and technological advancements. Collaborative efforts among nations can lead to more ambitious missions and foster innovation that individual countries might struggle to achieve alone.

One of the most significant outcomes of international collaboration is the pooling of financial resources. Space missions often require substantial funding, which can be a barrier for many nations. By forming partnerships, countries can share the financial burden, making it feasible to undertake large-scale projects such as crewed missions to Mars or the establishment of bases on the Moon. For instance, the International Space Station (ISS) exemplifies how multiple nations can contribute funds, technology, and manpower to create a sustainable human presence in low Earth orbit, serving as a launch pad for deeper space exploration.

International collaboration enhances the sharing of scientific knowledge and expertise. Different countries bring unique strengths and capabilities to the table, leading to a more nuanced understanding of the challenges posed by space exploration. In fields such as , , and engineering, researchers from diverse backgrounds can collaborate on experiments and share findings, accelerating the pace of discovery. This synergy can improve the probability of success on individual missions, unify the global scientific community, and set the stage for future interplanetary initiatives.

Joint missions also promote global diplomacy and peace. The collaborative spirit of space exploration forges better relations between nations that might otherwise be at odds. By working together on missions to explore the Moon, Mars, or asteroids, countries can build trust and strengthen diplomatic ties. This cooperative framework is pertinent as humanity ventures further into the cosmos, as it cultivates a shared sense of responsibility for the preservation of space and the resources it offers. The peaceful exploration of celestial bodies can serve as a model for international relations on Earth.

Looking toward the future of interplanetary travel and exploration, the significance of international collaboration is extremely relevant. The challenges of sustaining life on other planets, developing advanced propulsion systems, and ensuring the safety and success of human missions require a concerted effort from the global community. By embracing a collaborative approach, humanity can expand its reach beyond the blue planet, maximizing the potential for discovery and establishing a new era of exploration that unites us in our quest for knowledge and understanding of the universe.

Ethical Considerations in Space Exploration

Ethical considerations in space exploration encompass a range of issues that must be addressed as humanity ventures further into the cosmos. As explorers intending to traverse the Solar System and beyond, it is imperative to reflect on the implications of our activities on other celestial bodies. The potential for contamination of pristine environments raises significant moral questions about our responsibility to preserve extraterrestrial ecosystems. The introduction of Earth-based organisms could disrupt local environments and alter the natural progression of any potential life forms, whether simple microbial life or more complex organisms.

Another key ethical consideration involves the rights of potential . If we encounter intelligent life forms during our explorations, we must grapple with the ethical implications of our interactions. Should we impose our values and governance upon them, or should we adopt a more respectful approach that acknowledges their autonomy and cultural significance? Such dilemmas compel us to think critically about the principles that should guide our conduct in the vast expanse of the Milky Way and beyond, ensuring that our explorations do not lead to the subjugation or exploitation of other sentient beings.

The question of resource utilization also poses ethical challenges. As we consider mining asteroids or extracting materials from other planets for use on Earth, we must evaluate the sustainability of such actions. The potential depletion of resources from celestial bodies raises concerns about whether we are acting as responsible stewards of the universe or merely extending our consumption patterns into space. This consideration calls for the development of robust frameworks and agreements among nations to regulate resource extraction and ensure that future generations can also benefit from the riches of the Solar System.

The pursuit of interplanetary travel raises concerns about the impact on human health and the psychological well-being of explorers.

Long-duration missions can lead to isolation, confinement, and the stress of uncertainty, which can significantly affect the mental health of crew members. Ethical considerations must include the provision of support systems and safeguards to protect the well-being of those who venture into the unknown. As we prepare to embark on these journeys, we must prioritize the welfare of the explorers who will pave the way for humanity's presence among the stars.

Lastly, the issue of presents an ethical dilemma that requires urgent attention. As more missions are launched and satellites are deployed, the accumulation of debris poses risks not only to current space operations but also to future explorations. We must consider the long-term implications of our actions in orbit and adopt strategies to mitigate the creation of . This responsibility extends beyond our immediate needs and reflects our commitment to preserving the integrity of space for future explorers who will follow in our footsteps. By addressing these ethical considerations, we can ensure that our quest for knowledge and exploration is guided by principles that honor both our planet and the universe at large.

Chapter 14: Preparing for the Journey

Training for Interplanetary Travel

Training for interplanetary travel involves comprehensive preparation that goes beyond mere physical fitness. Explorers must undergo rigorous physical, psychological, and technical training to ensure they can withstand the unique challenges posed by space travel. The physical aspect includes strength and endurance training tailored to simulate the effects of microgravity and high-radiation environments. Explorers participate in exercise regimens designed to maintain muscle mass and bone density, crucial for long missions where the absence of Earth's gravity can lead to significant health issues.

 is equally important in interplanetary travel. The isolation and confinement of space missions can take a toll on mental health. Explorers engage in training that includes simulations of extended missions in isolated environments, helping them to develop coping strategies for stress and interpersonal conflict. Techniques such as mindfulness, team-building exercises, and conflict resolution training are incorporated to prepare explorers for the social dynamics that will unfold during their journeys. A thorough knowledge of the psychological aspects of long-duration spaceflight is central to maintaining team cohesion and individual well-being.

 is another cornerstone of training for interplanetary explorers. Participants undergo extensive education on spacecraft systems, navigation, and emergency procedures. This training often includes hands-on experience with simulators that replicate the equipment and technology they will encounter during their missions. Explorers learn to troubleshoot potential malfunctions and conduct repairs in a zero-gravity environment, ensuring they are prepared for any technical challenges that may arise during their travels through the solar system.

In addition to physical and psychological training, explorers must also familiarize themselves with the scientific objectives of their missions. This includes understanding the geology, atmospheres, and potential life forms on other celestial bodies. Training programs often involve collaborations with scientists and engineers to develop a comprehensive understanding of the mission's goals. By engaging in this interdisciplinary approach, explorers can magnify their scientific knowledge and develop a deeper appreciation for the environments they will explore.

Cultural and ethical training is becoming increasingly relevant as humanity expands its reach into the solar system. Explorers are educated on the implications of their presence on other planets, including the preservation of extraterrestrial environments and the importance of respecting potential indigenous life forms. This aspect of training promotes a sense of responsibility and stewardship among explorers, guiding them to conduct their missions in a manner that honors both the scientific and ethical dimensions of interplanetary exploration. With humanity preparing to venture beyond Earth, such training is essential to ensure that the legacy of exploration is one of curiosity tempered by respect and care for the cosmos.

Psychological Aspects of Space Travel

Psychological aspects of space travel encompass a wide range of factors that can significantly impact astronauts during their missions. One of the most prominent concerns is the psychological strain caused by isolation and confinement. Astronauts are often required to live and work in limited spaces for extended periods, which can lead to feelings of loneliness, anxiety, and even depression. The unique environment of space, combined with the absence of familiar social interactions, creates a setting that can amplify existing psychological issues or even lead to new ones. Addressing these challenges is required to ensure the mental well-being of explorers venturing beyond Earth.

Another major aspect is the impact of microgravity on psychological health. The physical effects of weightlessness can influence mood and cognitive function, as the body experiences a range of changes that may not be conducive to optimal mental performance. Research indicates that prolonged exposure to microgravity can lead to alterations in sleep patterns and circadian rhythms, which are vital for maintaining psychological stability. Preparations for interplanetary journeys, must take into account these effects and develop strategies to mitigate potential negative outcomes.

Social dynamics among crew members is also a high priority in the psychological well-being of astronauts. The crew's ability to communicate effectively, resolve conflicts, and support one another is paramount for mission success. In confined environments, interpersonal relationships can become strained, leading to misunderstandings and tension. Training programs that focus on teamwork and conflict resolution are essential to prepare explorers for the unique challenges they will face in space. Establishing solid social bonds can help alleviate stress and encourage a cooperative atmosphere critical for long-duration missions.

Coping mechanisms are another area of focus when discussing the psychological aspects of space travel. Astronauts are trained to develop strategies to manage stress and maintain mental health during their missions. These strategies may include mindfulness practices, physical exercise, and engaging in leisure activities to promote relaxation and cognitive engagement. The availability of psychological support, such as access to mental health professionals and communication with loved ones back on Earth, can also play a significant role in helping astronauts cope with the psychological demands of space travel.

While pursuing our goal of interplanetary exploration, acknowledging the psychological aspects of space travel must be given serious consideration. Innovations in spacecraft design, mission planning, and crew selection will need to incorporate considerations for mental health. By prioritizing psychological well-being, we can strengthen the resilience and improve the performance of explorers as they embark on journeys into the depths of the Milky Way. This focus on mental health will improve the chances of mission success and the overall experience of those who dare to venture beyond our blue planet.

Planning an Interplanetary Mission

Planning an interplanetary mission involves a meticulous and multi-faceted approach that integrates scientific objectives, technological advancements, and logistical considerations. The first step in this process is defining the mission's goals, which may range from conducting scientific research to testing new technologies for future exploration. This requires collaboration among scientists, engineers, and mission planners to establish clear objectives that align with the mission's capabilities and the specific characteristics of the target planet or moon.

Once the objectives are set, the next phase involves selecting the appropriate spacecraft and instruments. Different celestial bodies present unique challenges and opportunities, necessitating tailored designs that can withstand harsh environments, such as extreme temperatures, radiation, and dust storms. Engineers must consider propulsion systems, power sources, and communication technologies that will enable the spacecraft to travel vast distances and operate effectively upon arrival. Advances in materials science and robotics play a principle role in developing instruments that are not only durable but also capable of conducting complex experiments.

The trajectory of the mission is another primary aspect of planning. Calculating the optimal path to the target involves understanding planetary alignments, gravitational assists, and the timing of launch windows to minimize travel time and energy consumption. Mission planners use sophisticated software to model various scenarios, ensuring that the spacecraft will arrive at the intended destination with the necessary resources for a successful operation. This phase also encompasses contingency planning to address potential challenges that may arise during the journey.

Budgeting and funding are essential components of mission planning. Securing financial support from governmental and private sources can dictate the scope and scale of the mission. Detailed cost assessments must account for development, launch, operation, and potential contingencies. Engaging stakeholders and the public is also beneficial, as generating interest in interplanetary exploration can lead to increased investment and support. Transparency in budgeting processes helps build trust and stimulates collaboration among various entities involved in the mission.

Finally, the launch and operational phases require precise coordination among various teams to ensure mission success. A well-orchestrated launch sequence involves timing, resource allocation, and communication across multiple platforms. Once in space, continuous monitoring and adjustments may be necessary to account for unforeseen circumstances. The culmination of years of planning, research, and collaboration, an interplanetary mission exemplifies the spirit of exploration, pushing the boundaries of human knowledge and capability in the quest to understand our solar system and beyond.

The Importance of Continued Exploration

The importance of continued exploration of our solar system is a noble pursuit, as it serves as a cornerstone for advancing our understanding of the universe and our place within it. Exploration encourages scientific discovery, revealing fundamental truths about planetary formation, the potential for life beyond Earth, and the intricate dynamics of celestial bodies. Each mission into the depths of space acts as a magnifying glass, allowing us to scrutinize the complexities of the cosmos. The knowledge gained from these endeavors helps to unite our scientific communities and inspires future generations to engage with the mysteries of the universe.

Following the urge to journey beyond our home planet has the result of accelerating the technological advancements necessary for interplanetary travel. Continued exploration drives innovation in various fields, including engineering, robotics, and materials science. The challenges of traveling to Mars, the moons of Jupiter, or the rings of Saturn compel researchers to develop new technologies that can withstand extreme conditions, enhance navigation, and safeguard human life. These innovations often find applications on Earth, improving our daily lives and addressing pressing societal challenges, thereby highlighting the interconnectedness of space exploration and terrestrial advancements.

The quest to discover extraterrestrial life remains a compelling motivator for exploration. The search for microbial life on Mars or the potential subsurface oceans of Europa is exciting for scientists and the public alike. Finding out whether or not alien life exists elsewhere in the Milky Way Galaxy can reshape our philosophical and existential perspectives. Each discovery, whether confirming or refuting the possibility of life beyond Earth, adds depth to our understanding of biology, evolution, and the conditions necessary for life to thrive. The implications of such findings could redefine our place in the universe and influence future explorations.

Sustained exploration also plays a commanding role in planetary defense. As our ability to detect potentially hazardous asteroids and comets improves, so does our responsibility to protect our planet from cosmic threats. Continued missions to monitor these celestial objects and understand their trajectories are essential for developing strategies to mitigate risks. This proactive approach can help to safeguard our planet while at the same time enhancing our capabilities for future interplanetary travel and colonization, ensuring that humanity can endure and thrive in the face of cosmic challenges.

The dream of interplanetary travel captures the imagination, sparking curiosity and creativity in individuals across the globe. As explorers, we have the opportunity to unite diverse communities around a common goal — the pursuit of knowledge and discovery. It is for these reasons and more that the cultural and inspirational aspects of continued exploration cannot be overlooked. The stories of astronauts, rovers, and missions inspire collaboration and encourage a sense of global citizenship, reminding us that we are part of a larger cosmic narrative. By embracing the importance of continued exploration, we open the door to new possibilities, promoting a spirit of adventure that transcends boundaries and propels humanity toward a brighter future among the stars.

The Legacy of Interplanetary Travel

The legacy of interplanetary travel is a testament to human ingenuity and the relentless pursuit of knowledge. Since the dawn of the space age, the dream of traversing the vast expanse of our solar system has captivated the imagination of explorers and scientists. From the early missions that sent robotic probes to Venus and Mars to the ambitious plans for crewed missions to the outer planets, each step taken has laid a foundation for future exploration. The technological advancements and discoveries have expanded our understanding of our celestial neighbors and inspired generations to look towards the stars.

One of the most significant impacts of interplanetary travel has been the wealth of scientific data collected from various missions. Spacecraft such as Voyager, Cassini, and New Horizons have provided invaluable insights into the atmospheres, surfaces, and potential habitability of other planets and moons. These missions have revealed the complexity of our solar system, showcasing the diversity of planetary systems and the potential for life beyond Earth. The legacy of these explorations is not merely academic; it shapes the way we view our place in the universe and informs our understanding of Earth's own climate and geology through comparative analysis.

The technological innovations born from interplanetary travel have cascaded into numerous industries on Earth. Developments in materials science, propulsion technologies, and robotics have found applications beyond space exploration, influencing fields such as telecommunications, healthcare, and environmental monitoring. The pursuit of interplanetary travel has driven advancements that benefit society as a whole, proving that the quest for knowledge can yield practical solutions to terrestrial challenges. This dual legacy, both scientific and technological, underscores the importance of continued investment in space exploration.

There is also a significant cultural impact of interplanetary travel. As humanity reaches beyond its home planet, it produces a collective consciousness that transcends national borders. The images of distant worlds and the stories of exploration resonate with people from all walks of life, igniting a passion for science and discovery. The shared experience of rooting for missions like the Mars rovers or the landing on comets inspires a sense of unity and motivates future generations to consider careers in science, technology, engineering, and mathematics (). The legacy of these explorations serves as a reminder of what can be achieved when humanity comes together with a common purpose.

Looking ahead, the legacy of interplanetary travel is poised to evolve as new missions are planned and technologies are developed. With ambitions to establish a human presence on Mars and explore the icy moons of Jupiter and Saturn, future explorations will build upon the achievements of the past. As we venture deeper into the solar system, the lessons learned from previous missions will guide us in overcoming the challenges that lie ahead. The legacy of interplanetary travel will continue to shape our understanding of the cosmos, inspire future explorers, and remind us of the boundless possibilities that await in the vastness of space.

Inspiring Future Generations of Explorers

Inspiring future generations of explorers begins with instilling a sense of wonder about the cosmos. The Solar System is not merely a collection of celestial bodies; it is a vast realm filled with mysteries waiting to be uncovered. By seeding curiosity in young minds, we can ignite a passion for exploration that transcends generations. Educational programs that emphasize hands-on experiences, such as interactive planetarium shows, science fairs, and workshops, can engage students in a way that textbooks alone cannot. These experiences allow aspiring explorers to visualize their place in the universe and understand the importance of inquiry and discovery.

The stories of past explorers serve as powerful motivators for those who will venture into the cosmos in the future. The achievements of pioneers like Galileo, , and more recently, the teams behind the Mars rovers and Voyager missions, exemplify the spirit of human curiosity and perseverance. Sharing these narratives in schools and community programs can create role models for young explorers. When children learn about the challenges and triumphs faced by those who came before them, they can see a pathway to their own future in interplanetary travel and exploration, inspiring them to dream big and pursue careers in science, technology, engineering, and mathematics (STEM).

Advancements in technology have made space more accessible than ever, and this democratization of exploration can inspire future generations. Initiatives such as , , and international collaborations are paving the way for a new wave of explorers. By highlighting these opportunities, we can encourage young people to participate in space-related activities, whether through internships, competitions, or community-based projects. Engaging with technology that allows for real-time data sharing from space missions can also help demystify the exploration process, making it relatable and attainable for aspiring explorers.

Creating a culture of collaboration and communication among diverse communities can enhance the exploration experience. By encouraging teamwork across disciplines and cultures, we can harness a broader array of perspectives and ideas. Programs that promote inclusivity in STEM fields can help ensure that the next generation of explorers is as diverse as the Solar System itself. This diversity can lead to innovative solutions to the challenges of space exploration and promote a sense of global citizenship, where young explorers see themselves as part of a larger mission that transcends borders.

Finally, the importance of environmental stewardship cannot be ignored in the context of inspiring future generations. As we explore other planets and moons, we must also care for our home planet. Educational campaigns that emphasize the interconnectedness of Earth and the cosmos can cultivate a sense of responsibility among young explorers. By understanding that their actions have implications beyond our planet, they can grow into advocates for sustainable practices both on Earth and in their future explorations. This holistic approach will ensure that as future generations embark on their journeys into the Milky Way, they do so with a deep respect for the universe and a commitment to preserving the beauty of our own blue planet.

www.ingramcontent.com/pod-product-compliance
Lightning Source LLC
Chambersburg PA
CBHW071107240526
45469CB00006BD/2362